INVISIBLE WOMEN

BY THE SAME AUTHOR

Do it Like a Woman

INVISIBLE WOMEN

Data Bias in a World Designed for Men

Caroline Criado Perez

ABRAMS PRESS, NEW YORK

Library of Congress Control Number: 2019935419

Paperback ISBN: 978-1-4197-3521-9
eISBN: 978-1-68335-314-0

Printed and bound in the United States
11

ABRAMS The Art of Books
195 Broadway, New York, NY 10007
abramsbooks.com

For the women who persist: keep on being bloody difficult

Contents

Representation of the world, like the world itself, is the work of men; they describe it from their own point of view, which they confuse with the absolute truth.

Simone de Beauvoir

Preface

Most of recorded human history is one big data gap. Starting with the theory of Man the Hunter, the chroniclers of the past have left little space for women's role in the evolution of humanity, whether cultural or biological. Instead, the lives of men have been taken to represent those of humans overall. When it comes to the lives of the other half of humanity, there is often nothing but silence.

And these silences are everywhere. Our entire culture is riddled with them. Films, news, literature, science, city planning, economics. The stories we tell ourselves about our past, present and future. They are all marked – disfigured – by a female-shaped 'absent presence'. This is the gender data gap.

The gender data gap isn't just about silence. These silences, these gaps, have consequences. They impact on women's lives every day. The impact can be relatively minor. Shivering in offices set to a male temperature norm, for example, or struggling to reach a top shelf set at a male height norm. Irritating, certainly. Unjust, undoubtedly.

But not life-threatening. Not like crashing in a car whose safety measures don't account for women's measurements. Not like having your heart attack go undiagnosed because your symptoms are deemed 'atypical'. For these women, the consequences of living in a world built around male data can be deadly.

One of the most important things to say about the gender data gap is that it is not generally malicious, or even deliberate. Quite the opposite. It is simply the product of a way of thinking that has been around for millennia and is therefore a kind of *not* thinking. A double not thinking, even: men go without saying, and women don't get said at all. Because when we say human, on the whole, we mean man.

This is not a new observation. Simone de Beauvoir made it most famously when in 1949 she wrote, 'humanity is male and man defines woman not in herself, but as relative to him; she is not regarded as an autonomous being. [...] He is the Subject, he is the Absolute – she is the Other.'[1] What is new is the context in which women continue to be 'the Other'. And that context is a world increasingly reliant on and in thrall to data. Big Data. Which in turn is panned for Big Truths by Big Algorithms, using Big Computers. But when your big data is corrupted by big silences, the truths you get are half-truths, at best. And often, for women, they aren't true at all. As computer scientists themselves say: 'Garbage in, garbage out.'

This new context makes the need to close the gender data gap ever more urgent. Artificial intelligence that helps doctors with diagnoses, that scans through CVs, even that conducts interviews with potential job applicants, is already common. But AIs have been trained on data sets that are riddled with data gaps – and because algorithms are often protected as proprietary software, we can't even examine whether these gaps have been taken into account. On the available evidence, however, it certainly doesn't look as if they have.

Numbers, technology, algorithms, all of these are crucial to the story of *Invisible Women*. But they only tell half the story. Data is just another word for information, and information has many sources. Statistics are a kind of information, yes, but so is human experience. And so I will argue that when we are designing a world that is meant to work for everyone we need women in the room. If the people taking decisions that affect all of us are all white, able-bodied men (nine times out of ten from America), that too constitutes a data gap – in the same way that not collecting information on female bodies in medical research is a data gap. And as I will show, failing to include the perspective of women is a huge driver of an unintended male bias that attempts (often in good faith) to pass itself off as 'gender neutral'. This is what de Beauvoir meant when she said that men confuse their own point of view with the absolute truth.

The female-specific concerns that men fail to factor in cover a wide variety of areas, but as you read you will notice that three themes crop up again and again: the female body, women's unpaid care burden, and male violence against women. These are issues of such significance that they touch on nearly every part of our lives, affecting our experiences of everything from public transport to politics, via the workplace and the doctor's surgery. But men forget them, because men do not have female bodies. They, as we will see, do only a fraction of the unpaid work done by women. And while they do have to contend with male violence, it manifests in a different way to the violence faced by women. And so these differences go ignored, and we proceed as if the male body and its attendant life experience are gender neutral. This is a form of discrimination against women.

Throughout this book I will refer to both sex and gender. By 'sex', I mean the biological characteristics that determine whether an individual is male or female. XX and XY. By 'gender', I mean the social meanings we impose upon those biological facts – the way women are treated because they are *perceived* to be female. One is man-made, but both are real. And both have significant consequences for women as they navigate this world constructed on male data.

But although I talk about both sex and gender throughout, I use *gender* data gap as an overarching term because sex is not the reason women are excluded from data. Gender is. In naming the phenomenon that is causing so much damage to so many women's lives, I want to be clear about the root cause and, contrary to many claims you will read in these pages, the female body is not the problem. The problem is the social meaning that we ascribe to that body, and a socially determined failure to account for it.

Invisible Women is a story about absence – and that sometimes makes it hard to write about. If there is a data gap for women overall (both because we don't collect the data in the first place and because when we do we usually don't separate it by sex), when it comes to women of colour, disabled women, working-class women, the data is practically non-existent. Not simply because it isn't collected, but because it is not separated out from the male data – what is called 'sex-disaggregated data'. In statistics on representation from academic jobs to film roles, data is given for 'women' and 'ethnic minorities', with data for female ethnic minorities lost within each larger group. Where they exist, I have given them – but they barely ever do.

The point of this book is not psychoanalysis. I do not have direct access to the innermost thoughts of those who perpetuate the gender data gap, which means that this book cannot provide ultimate proof for why the gender data gap exists. I can only present you with the data, and ask you as a reader to look at the evidence. But nor am I interested in whether or not the person who produced a male-biased tool was a secret sexist. Private motivations are, to a certain extent, irrelevant. What matters is the pattern. What matters is whether, given the weight of the data I will present, it is reasonable to conclude that the gender data gap is all just one big coincidence.

I will argue that it is not. I will argue that the gender data gap is both a cause and a consequence of the type of unthinking that conceives of humanity as almost exclusively male. I will show how often and how widely this bias crops up, and how it distorts the supposedly objective data that increasingly rules our lives. I will show that even in this super-rational world increasingly run by super-impartial supercomputers, women are still very much de Beauvoir's *Second Sex* – and that the dangers of being relegated to, at best, a sub-type of men, are as real as they have ever been.

Introduction: The Default Male

Seeing men as the human default is fundamental to the structure of human society. It's an old habit and it runs deep – as deep as theories of human evolution itself. In the fourth century BC Aristotle was already baldly articulating male default as unarguable fact: 'The first departure from type is indeed that the offspring should become female instead of male', he wrote in his biological treatise *On the Generation of Animals.* (He did allow that this aberration was, however, 'a natural necessity'.)

Over two thousand years later, in 1966, the University of Chicago held a symposium on primitive hunter-gatherer societies. It was called 'Man the Hunter'. Over seventy-five social anthropologists from around the world gathered to debate the centrality of hunting to human evolution and development. The consensus was that it is pretty central.[1] 'The biology, psychology, and customs that separate us from the apes – all these we owe to the hunters of time past', claimed one of the papers published in the resulting book. Which is all very well, only, as feminists pointed out, this theory poses something of a problem for female evolution. Because, as the book made clear, hunting was a male activity. So if 'our intellect, interests, emotions, and basic social life – all are evolutionary products of the success of hunting adaptation', what does that mean for women's humanity? If human evolution is driven by men, are women even human?

In her now classic 1975 essay, 'Woman the Gatherer', anthropologist Sally Slocum challenged the primacy of 'Man the Hunter'.[2] Anthropologists, she argued, 'search for examples of the behaviour of males and assume that this is sufficient for explanation'. And so she asked a simple question to fill the silence: 'what were the females doing while the males were out hunting?' Answer: gathering, weaning, caring for children during 'longer periods of infant dependency', all of which would similarly have required cooperation. In the context of this knowledge, the 'conclusion that the basic human adaptation was the desire of males to hunt and kill,' objects Slocum, 'gives too much importance to aggression, which is after all only one factor of human life.'

Slocum made her critique over forty years ago now, but the male bias in evolutionary theory persists. 'Humans evolved to have an instinct for deadly violence, researchers find', read a 2016 headline in the *Independent*.[3] The article reported on an academic paper called 'The phylogenetic roots of human lethal violence', which claimed to reveal that humans have evolved to be six times more deadly to their own species than the average mammal.[4]

This is no doubt true of our species overall – but the reality of human-on-human lethal violence is that it is overwhelmingly a male occupation: a thirty-year analysis of murder in Sweden found that nine out of ten murders are committed by men.[5] This holds with statistics from other countries, including Australia,[6] the UK[7] and the US.[8] A 2013 UN homicide survey found that 96%[9] of homicide perpetrators worldwide are male. So is it humans who are murderous, or men? And if women aren't on the whole murdering, what are we to think of female 'phylogenetics'?

The male-unless-otherwise-indicated approach to research seems to have infected all sorts of ethnographic fields. Cave paintings, for example, are often of game animals and so researchers have assumed they were done by men – the hunters. But new analysis of handprints that appear alongside such paintings in cave sites in France and Spain has suggested that the majority were actually done by women.[10]

Even human bones are not exempt from male-unless-otherwise-indicated thinking. We might think of human skeletons as being objectively either male or female and therefore exempt from male-default thinking. We would be wrong. For over a hundred years, a tenth-century Viking skeleton known as the 'Birka warrior' had – despite possessing an apparently female pelvis – been assumed to be male because it was buried alongside a full set of weapons and two sacrificed horses.[11] These grave contents indicated that the occupant had been a warrior[12] – and warrior meant male (archaeologists put the numerous references to female fighters in Viking lore down to 'mythical embellishments'[13]). But although weapons apparently trump the pelvis when it comes to sex, they don't trump DNA and in 2017 testing confirmed that these bones did indeed belong to a woman.

The argument didn't, however, end there. It just shifted.[14] The bones might have been mixed up; there might be other reasons a female body was buried with these items. Naysaying scholars might have a point on both counts (although based on the layout of the grave contents the original authors dismiss these criticisms). But the resistance is nevertheless revealing, particularly since male skeletons in similar circumstances 'are not questioned in the same way'.[15] Indeed, when archaeologists dig up grave sites, they nearly always find more males, which, as noted anthropologist Phillip

Walker drily noted in a 1995 book chapter on sexing skulls, is 'not consistent with what we know about the sex ratios of extant human populations'.[16] And given Viking women could own property, could inherit and could become powerful merchants, is it so impossible that they could have fought too?[17]

After all, these are far from the only female warrior bones that have been discovered. 'Battle-scarred skeletons of multiple women have been found across the Eurasian steppes from Bulgaria to Mongolia' wrote Natalie Haynes in the *Guardian*.[18] For people such as the ancient Scythians, who fought on horseback with bows and arrows, there was no innate male warrior advantage, and DNA testing of skeletons buried with weapons in more than 1,000 Scythian burial mounds from Ukraine to Central Asia have revealed that up to 37% of Scythian women and girls were active warriors.[19]

The extent to which male-unless-otherwise-indicated permeates our thinking may seem less surprising when you realise that it is also embedded in one of the most basic building blocks of society: language itself. Indeed, when Slocum criticised male bias in anthropology, she pointed out that this bias appeared 'not only in the ways in which the scanty data are interpreted, but in the very language used'. The word 'man', she wrote, 'is used in such an ambiguous fashion that it is impossible to decide whether it refers to males or to the human species in general'. This collapse in meaning led Slocum to suspect that 'in the minds of many anthropologists, 'man', supposedly meaning the human species, is actually exactly synonymous with 'males'. As we shall see, the evidence suggests that she was probably right.

In Muriel Rukeyser's poem 'Myth', an old, blind Oedipus asks the Sphinx, 'Why didn't I recognize my mother?' The Sphinx replies that Oedipus answered her question (what walks on four legs in the morning, two in the afternoon and three in the evening) incorrectly. '[Y]ou answered, Man. You didn't say anything about woman.' But, replies Oedipus, when you say man, 'you include women too. Everyone knows that.'

But in fact the Sphinx was right and Oedipus is wrong. When you say man you don't 'include women too', even if everyone *does* technically 'know that'. Numerous studies in a variety of languages over the past forty years have consistently found that what is called the 'generic masculine' (using words like 'he' in a gender-neutral way) is not in fact read generically.[20] It is read overwhelmingly as male.

When the generic masculine is used people are more likely to recall famous men than famous women;[21] to estimate a profession as male-dominated;[22] to suggest male candidates for jobs and political appointments.[23] Women are also less likely to apply, and less likely to perform well in interviews, for jobs that are advertised using the generic masculine.[24] In fact the generic masculine is read so overwhelmingly as male that it even overrides otherwise powerful stereotypes, so that professions such as 'beautician', which are usually stereotyped female, are suddenly seen as male.[25] It even distorts scientific studies, creating a kind of meta gender data gap: a 2015 paper looking at self-report bias in psychological studies found that the use of the generic masculine in questionnaires affected women's responses, potentially distorting 'the meaning of test scores'.[26] The authors concluded that its use 'may portray unreal differences between women and men, which would not appear in the gender-neutral form or in natural gender language versions of the same questionnaire'.

And yet in the face of decades of evidence that the generic masculine is anything but clear, official language policy in many countries continues to insist that it is purely a formality whose use must continue for the sake of … clarity. As recently as 2017, the Académie française, France's ultimate authority on the French language, was thundering against 'the aberration of "inclusive writing"', claiming that 'the French language finds itself in mortal danger' from workarounds for the generic masculine. Other countries including Spain[27] and Israel[28] have faced similar rows.

Because English is not a grammatically gendered language, the generic masculine is fairly restricted in modern usage. Terms like 'doctor' and 'poet' used to be generic masculine (with specifically female doctors and poets referred to – usually derisively – as poetesses and doctoresses), but are now considered gender neutral. But while the formal use of the generic masculine only really clings on in the writings of pedants who still insist on using 'he' to mean 'he or she', it has made something of a comeback in the informal usage of Americanisms such as 'dude' and 'guys', and, in the UK, 'lads' as supposedly gender-neutral terms. A recent row in the UK also showed that, for some, male default still matters an awful lot: when in 2017 the first female head of London's Fire Brigade, Dany Cotton, suggested that we should replace 'fireman' with the now standard (and let's face it, much cooler) 'firefighter', she received a deluge of hate mail.[29]

Languages such as French, German and Spanish, however, are what is called 'gender-inflected', and here the concept of masculine and feminine is woven into the language itself. All nouns are gendered either masculine or feminine. A table is feminine, but a car is masculine: *la mesa roja* (the red table); *el coche rojo* (the red car). When it comes to nouns that refer to people, while both male and female terms exist, the standard gender is always masculine. Try searching Google for 'lawyer' in German. It comes back '*Anwalt*', which literally means male lawyer, but is also used generically as just 'lawyer'. If you want to refer to a female lawyer specifically you would say '*Anwältin*' (incidentally, the way female terms are often, as here, modified male terms is another subtle way we position the female as a deviation from male type – as, in de Beauvoir's terms, 'Other'). The generic masculine is also used when referring to groups of people: when the gender is unknown, or if it's a mixed group the generic masculine is used. So a group of one hundred female teachers in Spanish would be referred to as '*las profesoras*' – but as soon as you add a single male teacher, the group suddenly becomes '*los profesores*'. Such is the power of the default male.

In gender-inflected languages the generic masculine remains pervasive. Job vacancies are still often announced with masculine forms – particularly if they are for leadership roles.[30] A recent Austrian study of the language used in leadership jobs ads found a 27:1 ratio of masculine to 'gender-fair forms' (using both the male and female term).[31] The European Parliament believes it has found a solution to this problem, and since 2008 has recommended that '(m/f)' be added on the end of job ads in gender-inflected languages. The idea is that this makes the generic masculine more 'fair' by reminding us that women exist. It's a nice idea – but it wasn't backed up by data. When researchers *did* test its impact they found that it made no difference to the exclusionary impact of using the generic masculine on its own – illustrating the importance of collecting data and *then* creating policy.[32]

Does all this arguing over words make any real world difference? Arguably, yes. In 2012, a World Economic Forum analysis found that countries with gender-inflected languages, which have strong ideas of masculine and feminine present in almost every utterance, are the most unequal in terms of gender.[33] But here's an interesting quirk: countries with genderless languages (such as Hungarian and Finnish) are not the most equal. Instead, that honour belongs to a third group, countries with 'natural gender languages' such as English. These languages allow gender to be marked (female teacher, male nurse) but largely don't encode it into the words themselves. The study authors suggested that if you can't mark gender in any way you can't 'correct' the hidden bias in a language by emphasising 'women's presence in the world'. In short: because men go without saying, it matters when women literally can't get said at all.

It's tempting to think that the male bias that is embedded in language is simply a relic of more regressive times, but the evidence does not point that way. The world's 'fastest-growing language',[34] used by more than 90% of the world's online population, is emoji.[35] This language originated in Japan in the 1980s and women are its heaviest users:[36] 78% of women versus 60% of men frequently use emoji.[37] And yet, until 2016, the world of emojis was curiously male.

The emojis we have on our smartphones are chosen by the rather grand-sounding 'Unicode Consortium', a Silicon Valley-based group of organisations that work together to ensure universal, international software standards. If Unicode decides a particular emoji (say 'spy') should be added to the current stable, they will decide on the code that should be used. Each phone manufacturer (or platform such as Twitter and Facebook) will then design their own interpretation of what a 'spy' looks like. But they will all use the same code, so that when users communicate between different platforms, they are broadly all saying the same thing. An emoji face with heart eyes is an emoji face with heart eyes.

Unicode has not historically specified the gender for most emoji characters. The emoji that most platforms originally represented as a man running, was not called 'man running'. It was just called 'runner'. Similarly the original emoji for police officer was described by Unicode as 'police officer', not 'policeman'. It was the individual platforms that all interpreted these gender-neutral terms as male.

In 2016, Unicode decided to do something about this. Abandoning their previously 'neutral' gender stance, they decided to explicitly gender all emojis that depicted people.[38] So instead of 'runner' which had been universally represented as 'male runner', Unicode issued code for explicitly male runner and explicitly female runner. Male and female options now exist for all professions and athletes. It's a small victory, but a significant one.

It's easy to slam phone manufacturers and social media platforms as sexist (and, as we shall see, they are, if often unknowingly), but the reality is that even if they had somehow managed to design an image of a 'gender neutral' runner, most of us would still have read that runner as male, because we read most things as male unless they are specifically marked as female. And so while it is of course to be hoped that angry grammarians will come round to the idea that saying 'he and she' (or even, God forbid, 'she and he') instead of just 'he' may not be the worst thing that has ever happened to them, the truth is that getting rid of the generic masculine would only be half the battle: male bias is so firmly embedded in our psyche that even genuinely gender-neutral words are read as male.

A 2015 study identified the top five words used to refer to people in human–computer interaction papers published in 2014 and found that they are all apparently gender neutral: user, participant, person, designer and researcher.[39] Well done, human–computer interaction academics! But there is (of course) a catch. When study participants were instructed to think about one of these words for ten seconds and then draw an image of it, it turned out that these apparently gender-neutral words were not perceived as equally likely to be male or female. For male participants, only 'designer' was interpreted as male less than 80% of the time (it was still almost 70% male). A researcher was more likely to be depicted as of no gender than as a female. Women were slightly less gender-biased, but on the whole were still more likely to read gender-neutral words as male, with only 'person' and 'participant' (both read by about 80% of male participants as male) being about 50/50.

This rather disheartening finding tallies with decades of 'draw a scientist' data, where participants overwhelmingly draw men (the bias has historically been so extreme that media around the world celebrated as great progress a recent paper which found that 28%

of children now draw women).[40] It also tallies, perhaps more disturbingly, with a 2008 study in which Pakistani students (aged nine and ten) who were asked to draw an image of 'us'.[41] Hardly any of the female students drew women and none of the male students did.

We don't even allow non-humans to escape our perception of the world as overwhelmingly male: when researchers in one study attempted to prompt participants to see a gender-neutral stuffed animal as female by using female pronouns, children, parents and carers still overwhelmingly referred to the animal as 'he'.[42] The study found that an animal must be 'super-feminine' before 'even close to half of participants will refer to it as she rather than he'.

To be fair, it's not an entirely unreasonable assumption: often it really is a he. A 2007 international study of 25,439 children's TV characters found that only 13% of non-human characters are female (the figure for female human characters was slightly better, although still low at 32%).[43] An analysis of G-rated (suitable for children) films released between 1990 and 2005 found that only 28% of speaking roles went to female characters – and perhaps even more tellingly in the context of humans being male by default, women made up only 17% of crowd scenes.[44]

Men don't just have more roles, they also spend twice as much time on screen – this rises to nearly three times as much when, as most films do, the film has a male lead.[45] Only when the lead is female do men and women appear about as often as each other (as opposed to women getting, as you might expect, the majority of screen time). Men also get more lines, speaking twice as much as women overall; three times as much in films with male leads; and almost twice as much in films with male and female co-leads. Again it is only in the few films with female leads where male and female characters drew even on screen time.

This imbalance is found not just in films and TV. It's everywhere.

It's in statues: when I counted all the statues in the UK's Public Monuments and Sculptures Association database I found that there were more statues of men called John than there were of historical, named, non-royal women (the only reason adding royal women to the figure just beats the Johns is down to Queen Victoria, whose enthusiasm for putting up statues of herself I have a grudging respect for).

It's on banknotes: in 2013 the Bank of England announced they were replacing the only female historical figure on their banknotes with another man (I fought a successful campaign against it and campaigns have cropped up in other countries, including Canada and the US).[46]

It's in the news media: every five years since 1995, the Global Media Monitoring Project has evaluated the world's print and broadcast media for its representation of women. Its latest report, published in 2015, found that 'women make up only 24% of the persons heard, read about or seen in newspaper, television and radio news, exactly as they did in 2010'.[47]

It's even in school textbooks. Thirty years of language and grammar textbook studies in countries including Germany, the US, Australia, and Spain have found that men far outnumber women in example sentences (on average by about 3:1).[48] A US study of eighteen widely used high-school history textbooks published between 1960 and 1990 found that pictures of named men outnumbered pictures of named women by a ratio of about 18 to 100 and that only 9% of the names in the indexes were women (a figure that persisted into the 2002 edition of one of the textbooks).[49] More recently, a 2017 analysis of ten introductory political-science textbooks found that an average of only 10.8% of pages per text referenced women (some texts were as low as 5.3%).[50] The same level of male bias has been found in recent analyses of Armenian, Malawian, Pakistani, Taiwanese, South African and Russian textbooks.[51]

So widespread is this cultural bias towards representing men that the makers of the classic sci-fi action game series, *Metroid*, relied on it when they wanted to surprise their users. 'We wondered what would surprise everyone and talked about removing [main character] Samus's helmet. Then someone said, 'It would be a shocker if Samus turned out to be a woman!' they recalled in a recent interview.[52] And to make sure everyone really got it, they put her in a pink bikini and hip-jutting pose.

Metroid was – and remains – something of an outlier in gaming. Although a 2015 Pew Research Center report[53] found that equal numbers of American men and women play video games, only 3.3%[54] of the games spotlighted at press conferences during 2016's E3 (the world's largest annual gaming expo) starred female protagonists. This is actually lower than the figure for 2015 which, according to Feminist Frequency, was 9%.[55] If female playable characters do make it into a game they are still often framed as just another feature. At E3 2015 the director of *Fallout 4*, Todd Howard, revealed how easy it was to switch between male and female playable characters – only to switch back to the male version for the rest of the demo.[56] As Feminist Frequency remarked when they released their data on E3 2016, 'heroes are male by default'.[57]

The result of this deeply male-dominated culture is that the male experience, the male perspective, has come to be seen as universal, while the female experience – that of half the global population, after all – is seen as, well, niche. It is because what is male is universal that when a professor at Georgetown University named her literature course 'White Male Writers', she hit the headlines, while the numerous courses on 'female writers' pass unremarked.[58]

It is because what is male is universal (and what is female is niche) that a film about the fight of British women for their right to vote is slammed (in the *Guardian*, no less) as 'peculiarly hermetic' for not covering the First World War – sadly proving that Virginia Woolf's

1929 observation ('This is an important book, the critic assumes, because it deals with war. This is an insignificant book because it deals with the feelings of women in a drawing-room') is still relevant today.[59] It is why V. S. Naipaul criticises Jane Austen's writing as 'narrow', while at the same time no one is expecting *The Wolf of Wall Street* to address the Gulf War, or Norwegian writer Karl Ove Knausgaard to write about anyone but himself (or quote more than a single female writer) to receive praise from the *New Yorker* for voicing 'universal anxieties' in his six-volume autobiography.

It is why the England national football team page on Wikipedia is about the men's national football team, while the women's page is called the England women's national football team, and why in 2013 Wikipedia divided writers into 'American Novelists' and 'American Women Novelists'. It is why a 2015 study of multiple language Wikipedias found that articles about women include words like 'woman', 'female' or 'lady', but articles about men don't contain words like 'man', 'masculine' or 'gentleman' (because the male sex goes without saying).[60]

We class the fourteenth to seventeenth centuries as 'the Renaissance' even though, as social psychologist Carol Tavris points out in her 1991 book *The Mismeasure of Woman*, it wasn't a renaissance for women, who were still largely excluded from intellectual and artistic life. We call the eighteenth century 'the Enlightenment', even though, while it may have expanded 'the rights of man', it 'narrowed the rights of women, who were denied control of their property and earnings and barred from higher education and professional training'. We think of ancient Greece as the cradle of democracy although the female half of the population were explicitly excluded from voting.

In 2013, British tennis player Andy Murray was lauded across the media for ending Britain's '77-year wait' to win Wimbledon, when in fact Virginia Wade had won it in 1977. Three years later,

Murray was informed by a sports reporter that he was 'the first person ever to win two Olympic tennis gold medals' (Murray correctly replied that 'Venus and Serena have won about four each').[61] In the US it is a truth universally acknowledged that its soccer team has never won the World Cup or even reached the final – except it has. Its women's team has won four times.[62]

Recent years have seen some laudable attempts to address this relentless male cultural bias, but these are often met with hostility. When Thor was reinvented as a woman by Marvel Comics,[63] fans revolted – although as *Wired* magazine pointed out, 'no one uttered a peep' when Thor was replaced by a frog.[64] When the *Star Wars* franchise released two films in a row with a female lead howls of outrage reverberated around the manosphere.[65] One of the UK's longest-running television shows (*Doctor Who*) is a sci-fi fantasy series about a shape-shifting alien who periodically morphs into a new body, and the alien's first twelve incarnations were all male. But in 2017, for the first time, the doctor morphed into a woman. In response, former doctor Peter Davison expressed 'doubts' about the wisdom of casting a woman in the role of Doctor Who.[66] He preferred the idea of the doctor as 'a boy' and mourned 'the loss of a role model for boys'. Upset men took to Twitter calling for a boycott of the show, condemning the decision as 'PC' and 'liberal' virtue-signalling.[67]

Colin Baker, the body into whom the Peter Davison doctor had morphed, disagreed with his predecessor. Boys have 'had fifty years of having a role model', he argued. And in any case, he mused, do you have to be the same gender as someone to be a role model? 'Can't you be a role model as people?' Not really, Colin, because as we've seen, 'people' tends to be read as male. And in any case, while there is evidence that women can to a certain extent accept men as role models, men won't do the same for women. Women will buy books by and about men, but men won't buy books by and about women (or at least not many).[68] When adventure video game series *Assassin's Creed* announced in 2014 that it would not be possible to play as a female assassin in their new cooperative multiplayer mode, some male players were pleased with the decision.[69] Playing as a woman would alienate them from the game, they argued.

Journalist Sarah Ditum has little time for this argument. 'Come on now,' she chided in a column. 'You've played games as a blue hedgehog. As a cybernetically augmented space marine. As a sodding dragon-tamer. [...B]ut the idea that women can be protagonists with an inner life and an active nature is somehow beyond your imaginative capacities?'[70] Ditum is of course technically right. It *should* be easier to imagine yourself as a woman than as a blue hedgehog. But on the other hand she's also wrong, because that blue hedgehog has one particularly important similarity with male players, even more so than species alignment, and that is that Sonic the hedgehog is male. We know this because he isn't pink, he doesn't have a bow in his hair, and he doesn't simper. He is the standard, unmarked gender, not the atypical one.

This kind of negative reaction to the introduction of women is witnessed all over the cultural landscape. When in 2013 I campaigned to have a female historical figure on the back of English banknotes some men got so angry that they felt compelled to threaten me with rape, mutilation and death. Not all the men who disliked the campaign went that far, of course, but the sense of injustice was still clear in the more measured responses I got. I remember one man expostulating, 'but women are everywhere now!' Clearly, given I was having to campaign so hard for the inclusion of one woman, they aren't, but his perspective was nevertheless telling. These men were experiencing even minor female representation as an iniquity. As far as they were concerned, the playing field was already level, and the entirely male line-up was just an objective reflection of merit.

Before they caved, the Bank of England's case for their all-male line-up also rested on the meritocracy argument: historical figures were, they said, chosen using an 'objective selection criteria'. To join the 'gilded list' of 'key figures from our past', a person must fulfil the following: have broad name recognition; have good artwork; not be controversial; and have made 'a lasting contribution which is universally recognised and has enduring benefits'. Reading these subjective designations of worth, I realised how the Bank had ended up with five white men on its banknotes: the historical gender data gap means that women are just far less likely to be able to fulfil any of these 'objective' criteria.

In 1839 the composer Clara Schumann wrote in her diary, 'I once thought that I possessed creative talent, but I have given up this idea; a woman must not desire to compose – not one has been able to do it, and why should I expect to?' The tragedy is, Schumann was wrong. Women before her *had* been able to do it, and they included some of the most successful, prolific and influential composers of the seventeenth and eighteenth centuries.[71] It's just that they didn't have 'broad name recognition', because a woman barely has to die before she is forgotten – or before we consign her work to the gender data gap by attributing it to a man.

Felix Mendelssohn published six of his sister Fanny Hensel's pieces under his own name and in 2010 another manuscript previously thought to be his was proven to be Hensel's.[72] For years classical scholars argued that the Roman poet Sulpicia couldn't possibly have written the verses signed with her name – they were too good, not to mention too smutty.[73] Judith Leyster, one of the first Dutch women to be admitted to an artists' guild, was renowned in her time, but after her death in 1660 she was erased, her work attributed to her husband. In 2017, new works by nineteenth-century artist Caroline Louisa Daly were discovered – they had been previously attributed to men, one of whom was not even an artist.[74]

At the turn of the twentieth century, award-winning British engineer, physicist and inventor Hertha Ayrton remarked that while errors overall are 'notoriously hard to kill [...] an error that ascribes to a man what was actually the work of a woman has more lives than a cat'. She was right. Textbooks still routinely name Thomas Hunt Morgan as the person who discovered that sex was determined by chromosomes rather than environment, despite the fact that it was Nettie Stevens' experiments on mealworms that established this – and despite the existence of correspondence between them where Morgan writes to ask Stevens for details of her experiment.[75] Cecilia Payne-Gaposchkin's discovery that the sun is predominantly composed of hydrogen is often credited to her male supervisor.[76] Perhaps the most famous example of this kind of injustice is Rosalind Franklin, whose work (she had concluded via her X-ray experiments and unit cell measurements that DNA consisted of two chains and a phosphate backbone) led James Watson and Francis Crick (now Nobel Prize-winning household names) to 'discover' DNA.

None of this means that the Bank of England deliberately set out to exclude women. It just means that what may seem objective can actually be highly male-biased: in this case, the historically widespread practice of attributing women's work to men made it much harder for a woman to fulfil the Bank's requirements. The fact is that worth is a matter of opinion, and opinion is informed by culture. And if that culture is as male-biased as ours is, it can't help but be biased against women. By default.

The case of the Bank's subjective selection criteria also shows how male default can be both a cause and a consequence of the gender data gap. By neglecting to account for the historical gender data gap, the Bank's selection procedure for historical figures was designed around the kind of success typically achieved by men; even a requirement as seemingly benign as that the figure not be controversial, well, as the historian Laurel Thatcher Ulrich famously put it, 'well-behaved women seldom make history'. The result was that the Bank not only failed to correct for the historical gender data gap: it perpetuated it.

Such subjective designations of worth masquerading as objectivity crop up all over the place. In 2015 a British A level student called Jesse McCabe noticed that of the sixty-three set works included in her music syllabus, not a single one was by a woman. When she wrote to her exam board, Edexcel, they defended the syllabus. 'Given that female composers were not prominent in the western classical tradition (or others for that matter),' they wrote, 'there would be very few female composers that could be included.' The phrasing here is important. Edexcel doesn't mean that there simply aren't any female composers – after all, the *International Encyclopaedia of Women Composers* alone has more than 6,000 entries. What they are talking about here is 'the canon', that is, the body of works generally agreed to have been the most influential in shaping western culture.

Canon formation is passed off as the objective trickle-down of the musical marketplace, but in truth it is as subjective as any other value judgment made in an unequal society. Women have been locked out of the canon wholesale because what success looked like in composing has historically been almost impossible for women to achieve. For most of history, if women were allowed to compose at all, it was for a private audience and domestic setting. Large orchestral works, so crucial for the development of a composer's reputation, were usually off limits, considered 'improper'.[77] Music was an 'ornament' for women, not a career.[78] Even by the twentieth century, Elizabeth Maconchy (who was the first woman ever to chair the Composers' Guild of Great Britain), was being curtailed in her ambitions by publishers such as Leslie Boosey, who 'couldn't take anything except little songs from a woman'.

Even if the 'little songs' women were allowed to write were enough to earn you a place in the canon, women simply didn't have the resources or position to ensure their legacy. In her book *Sounds and Sweet Airs: The Forgotten Women of Classical Music*, Anna Beer compares the prolific seventeenth-century composer Barbara Strozzi (who 'had more music in print in her lifetime than any other composer of the era') to one of her male contemporaries, Francesco Cavalli. As head of music at St Mark's in Venice (a position not open to women at the time), Cavalli had the money and the stature to ensure all his works, including the many he did not publish in his lifetime, were kept in a library. He could pay for an archivist to look after them, and he could, and did, pay for the Masses he composed to be sung on the anniversary of his death. In the face of such inequality of resources, Strozzi never stood a chance of being remembered on an equal footing. And to continue to insist on the primacy of a canon that excludes women like her is to perpetuate the male-biased injustices of the past.

As well as going some way to explaining their exclusion from cultural history, the exclusion of women from positions of power is often given as an excuse for why, when we teach them about the past, we teach children almost exclusively about the lives of men. In 2013, a battle raged in Britain over what we mean by 'history'. On one side was the then British Secretary of State for Education, Michael Gove, brandishing his proposed new 'back to basics' national history curriculum.[79] An army of twenty-first-century Gradgrinds, he and his supporters insisted that children needed 'facts'.[80] They needed a 'foundation of knowledge'.

This 'foundation of knowledge', the 'basic' blocks of 'facts' which every child should know, was notable, amongst other gaps, for its almost wholesale absence of women. No women appeared in Key Stage 2 (ages seven to eleven) at all, other than two Tudor queens. Key Stage 3 (ages eleven to fourteen) included only five women, four of whom (Florence Nightingale, Mary Seacole, George Eliot and Annie Besant) were lumped together under 'The Changing Role of Women' – rather implying, not without reason, that the rest of the curriculum was about men.

In 2009, prominent British historian David Starkey criticised female historians for, in his opinion, focusing too much on Henry VIII's wives rather than the king himself who, he railed, should be 'centre stage'.[81] Dismissing the 'soap opera' of his personal life as secondary to the formal political consequences of his rule, such as the Reformation, Starkey insisted that '[i]f you are to do a proper history of Europe before the last five minutes it is a history of white males because they were the power players, and to pretend anything else is to falsify'.

Starkey's position rests on the assumption that what takes place in the private realm is unimportant. But is that a fact? The private life of Agnes Huntingdon (born after 1320) is revealed through snippets in public documents from the court cases concerning her two marriages.[82] We discover that she was a victim of domestic abuse, and that her first marriage was disputed because her family disapproved of her choice. On the evening of 25 July 1345 she ran away from her second husband after he attacked her; later that night he turned up at her brother's house with a knife. Is the abuse (and lack of freedom of choice) of a fourteenth-century woman private irrelevancies, or part of the history of female subjugation?

The arbitrary division of the world into 'private' and 'public' is in any case arguably a false distinction. Invariably both bleed into each other. When I spoke to Katherine Edwards, a history teacher who was heavily involved in the fight against Gove's reforms, she pointed to recent research on women's role in the American Civil War. Far from being an irrelevance, 'women and their conception of their own role completely undermined the whole Confederate war effort'.

Elite women, brought up to believe absolutely in the myth of their own helplessness, simply could not get over their understanding of work as intrinsically unfeminine. Unable to bring themselves to take up the jobs vacated by enlisted men, they wrote to their husbands begging them to desert, to come home and protect them. Poorer women proved a headache in a more proactive way, as they organised resistance to Confederate policies, 'because they were starving basically, and they needed to feed their families'. Excluding women from an analysis of the outcome of the American Civil War not only constitutes a gender data gap, but also a data gap in the understanding of the construction of the United States itself. That seems like a 'fact' worth knowing.

The history of humanity. The history of art, literature and music. The history of evolution itself. All have been presented to us as objective facts. But the reality is, these facts have been lying to us. They have all been distorted by a failure to account for half of humanity – not least by the very words we use to convey our half-truths. This failure has led to gaps in the data. A corruption in what we think we know about ourselves. It has fuelled the myth of male universality. And that *is* a fact.

The persistence of this myth continues to affect how we see ourselves today – and if the past few years have shown us anything it is that how we see ourselves is not a minor concern. Identity is a potent force that we ignore and misread at our peril: Trump, Brexit and ISIS (to name just three recent examples) are global phenomena that have upended the world order – and they are all, at heart, identity-driven projects. But misreading and ignoring identity is exactly what obfuscating maleness under the guise of gender-neutral universality causes us to do.

A man I briefly dated tried to win arguments with me by telling me I was blinded by ideology. I couldn't see the world objectively, he said, or rationally, because I was a feminist and I saw everything through feminist eyes. When I pointed out that this was true for him too (he identified as a libertarian) he demurred. No. That was just objective, common sense – de Beauvoir's 'absolute truth'. For him, the way he saw the world was universal, while feminism – seeing the world from a female perspective – was niche. Ideological.

I was reminded of this man in the wake of the 2016 US presidential election, when it felt you couldn't move for tweets, speeches and op-eds by (usually) white men decrying the ills of what they called 'identity politics'. Ten days after Donald Trump's victory, the *New York Times* published an article by Mark Lilla, professor of humanities at Columbia University, that criticised Clinton for 'calling out explicitly to African American, Latino, LGBT and women voters'.[83] This left out, he said, 'the white working class'. Lilla presented Clinton's 'rhetoric of diversity' as mutually exclusive with 'a large vision', linking this 'narrow' vision (clearly, Lilla has been reading his V. S. Naipaul) with what he felt he was witnessing with college students. Students today, he claimed, were so primed to focus on diversity that they 'have shockingly little to say about such perennial questions as class, war, the economy and the common good'.

Two days after this was published, ex-Democratic candidate Bernie Sanders was in Boston at a stop on his book tour[84] explaining that 'It is not good enough for someone to say, I'm a woman! Vote for me!'[85] In Australia, Paul Kelly, editor of the *Australian*, described Trump's victory as 'a revolt against identity politics',[86] while over in the UK, Labour MP Richard Burgon tweeted that Trump's inauguration was 'what can happen when centre/left parties abandon transformation of economic system and rely on identity politics'.[87]

The *Guardian*'s Simon Jenkins concluded the *annus horribilis* that was 2016 with a diatribe against 'the identity apostles', who had been 'over-defensive' of minorities, and thus killed off liberalism.

'I have no tribe,' he wrote. He could not 'join the prevailing hysteria'. What he wanted was 'to re-enact the glorious revolution of 1832' – which resulted in the extension of the British franchise to a few extra hundred thousand men of property.[88] Heady days, indeed.

These white men have in common the following opinions: that identity politics is only identity politics when it's about race or sex; that race and sex have nothing to do with 'wider' issues like 'the economy'; that it is 'narrow' to specifically address the concerns of female voters and voters of colour; and that working class means white working-class men. Incidentally, according to the US Bureau of Labor Statistics, the coal mining industry, which during the 2016 election became the shibboleth for (implicitly male) working-class jobs, provides 53,420 jobs in total, at a median annual wage of $59,380.[89] Compare this to the majority female 924,640-strong cleaning and housekeeper workforce, whose median annual income is $21,820.[90] So who's the real working class?

These white men also have in common that they are white men. And I labour this point because it is exactly their whiteness and maleness that caused them to seriously vocalise the logical absurdity that identities exist only for those who happen not to be white or male. When you have been so used, as a white man, to white and male going without saying, it's understandable that you might forget that white and male is an identity too.

Pierre Bourdieu wrote in 1977 that 'what is essential goes without saying because it comes without saying: the tradition is silent, not least about itself as a tradition'.[91] Whiteness and maleness are silent precisely because they do not need to be vocalised. Whiteness and maleness are implicit. They are unquestioned. They are the default. And this reality is inescapable for anyone whose identity does not go without saying, for anyone whose needs and perspective are routinely forgotten. For anyone who is used to jarring up against a world that has not been designed around them and their needs.

The way whiteness and maleness go without saying brings me back to my bad date (OK, dates), because it is intrinsically linked to the misguided belief in the objectivity, the rationality, the, as Catherine Mackinnon has it, 'point-of-viewlessness' of the white, male perspective. Because this perspective is not articulated as white and male (because it doesn't need to be), because it is the norm, it is presumed not to be subjective. It is presumed to be objective. Universal, even.

This presumption is unsound. The truth is that white and male is just as much an identity as black and female. One study which looked specifically at white Americans' attitudes and candidate preferences found that Trump's success reflected the rise of 'white identity politics', which the researchers defined as 'an attempt to protect the collective interests of white voters via the ballot box'.[92] White identity, they concluded, 'strongly predicts a preference for Trump'. And so did male identity. Analysis of how gender affected support for Trump revealed that 'the more hostile voters were toward women, the more likely they were to support Trump'.[93] In fact, hostile sexism was nearly as good at predicting support for Trump as party identification. And the only reason this is a surprise to us is because we are so used to the myth of male universality.

The presumption that what is male is universal is a direct consequence of the gender data gap. Whiteness and maleness can only go without saying because most other identities never get said at all. But male universality is also a *cause* of the gender data gap: because women aren't seen and aren't remembered, because male data makes up the majority of what we know, what is male *comes* to be seen as universal. It leads to the positioning of women, half the global population, as a minority. With a niche identity and a subjective point of view. In such a framing, women are set up to be forgettable. Ignorable. Dispensable – from culture, from history, from data. And so, women become invisible.

Invisible Women is the story of what happens when we forget to account for half of humanity. It is an exposé of how the gender data gap harms women when life proceeds, more or less as normal. In urban planning, politics, the workplace. It is also about what happens to women living in a world built on male data when things go wrong. When they get sick. When they lose their home in a flood. When they have to flee that home because of war.

But there is hope in this story too, because it's when women are able to step out from the shadows with their voices and their bodies that things start to shift. The gaps close. And so, at heart, *Invisible Women* is also a call for change. For too long we have positioned women as a deviation from standard humanity and this is why they have been allowed to become invisible. It's time for a change in perspective. It's time for women to be seen.

PART I

Daily Life

CHAPTER 1

Can Snow-Clearing be Sexist?

It all started with a joke. It was 2011 and officials in the town of Karlskoga, in Sweden, were being hit with a gender-equality initiative that meant they had to re-evaluate all their policies through a gendered lens. As one after another of their policies were subjected to this harsh glare, one unfortunate official laughed that at least snow-clearing was something the 'gender people' would keep their noses out of. Unfortunately for him, his comment got the gender people thinking: is snow-clearing sexist?

At the time, in line with most administrations, snow-clearing in Karlskoga began with the major traffic arteries, and ended with pedestrian walkways and bicycle paths. But this was affecting men and women differently because men and women travel differently.

We lack consistent, sex-disaggregated data from every country, but the data we do have makes it clear that women are invariably more likely than men to walk and take public transport.[1] In France, two-thirds of public transport passengers are women; in Philadelphia and Chicago in the US, the figure is 64%[2] and 62%[3] respectively. Meanwhile, men around the world are more likely to drive[4] and if a household owns a car, it is the men who dominate access to it[5] – even in the feminist utopia that is Sweden.[6]

And the differences don't stop at the mode of transport: it's also about *why* men and women are travelling. Men are most likely to have a fairly simple travel pattern: a twice-daily commute in and out of town. But women's travel patterns tend to be more complicated. Women do 75% of the world's unpaid care work and this affects their travel needs. A typical female travel pattern involves, for example, dropping children off at school before going to work; taking an elderly relative to the doctor and doing the grocery shopping on the way home. This is called 'trip-chaining', a travel pattern of several small interconnected trips that has been observed in women around the world.

In London women are three times more likely than men to take a child to school[7] and 25%[8] more likely to trip-chain; this figure rises to 39% if there is a child older than nine in the household. The disparity in male/female trip-chaining is found across Europe, where women in dual-worker families are twice as likely as men to pick up and drop off children at school during their commute. It is most pronounced in households with young children: a working woman with a child under the age of five will increase her trip-chaining by 54%; a working man in the same position will only increase his by 19%.[9]

What all these differences meant back in Karlskoga was that the apparently gender-neutral snow-clearing schedule was in fact not gender neutral at all, so the town councillors switched the order of snow-clearing to prioritise pedestrians and public-transport users. After all, they reasoned, it wouldn't cost any more money, and driving a car through three inches of snow is easier than pushing a buggy (or a wheelchair, or a bike) through three inches of snow.

What they didn't realise was that it would actually end up saving them money. Since 1985, northern Sweden has been collecting data on hospital admissions for injuries. Their databases are dominated by pedestrians, who are injured three times more often than motorists in slippery or icy conditions[10] and account for half the hospital time of all traffic-related injuries.[11] And the majority of these pedestrians are women. A study of pedestrian injuries in the Swedish city area of Umeå found that 79% occurred during the winter months, and that women made up 69% of those who had been injured in single-person incidents (that is, those which didn't involve anyone else). Two-thirds of injured pedestrians had slipped and fallen on icy or snowy surfaces, and 48% had moderate to serious injuries, with fractures and dislocations being the most common. Women's injuries also tended to be more severe.

A five-year study in Skåne County uncovered the same trends – and found that the injuries cost money in healthcare and lost productivity.[12] The estimated cost of all these pedestrian falls during just a single winter season was 36 million Kronor (around £3.2 million). (This is likely to be a conservative estimate: many injured pedestrians will visit hospitals that are not contributing to the national traffic accident register; some will visit doctors; and some will simply stay at home. As a result, both the healthcare and productivity costs are likely to be higher.)

But even with this conservative estimate, the cost of pedestrian accidents in icy conditions was about twice the cost of winter road maintenance. In Solna, near Stockholm, it was three times the cost, and some studies reveal it's even higher.[13] Whatever the exact disparity, it is clear that preventing injuries by prioritising pedestrians in the snow-clearing schedule makes economic sense.

A brief snow-clearing coda comes from the alt-right blogosphere,[14] which reacted with glee when Stockholm failed to execute a smooth transfer to gender-equal snow-clearing in 2016: an unusually high snowfall that year left roads and pavements covered in snow and commuters unable to get to work. But in their rush to celebrate the foundering of a feminist policy what these right-wing commentators failed to note was that this system had already been working successfully in Karlskoga for three years.

They also, in any case, reported the issue inaccurately. Heat St claimed[15] that the policy was a failure in part because 'injuries requiring a hospital visit reportedly spiked' – neglecting to note that it was pedestrian injuries[16] that had 'spiked', illustrating that the problem was not that pedestrians had been prioritised, but that snow-clearing as a whole had not been conducted effectively. Motorists may not have been travelling well, but neither was anyone else.

The following winter was much more successful: when I spoke to Daniel Helldén, a local councillor in Stockholm's traffic department, he told me that on the 200 km of joint cycle and pedestrian lanes that are now being cleared with special machines ('which make them as clean as in the summer') accidents have gone down by half. 'So it's a really good effect.'

The original snow-clearing schedule in Karlskoga hadn't been deliberately designed to benefit men at the expense of women. Like many of the examples in this book, it came about as a result of a gender data gap – in this instance, a gap in perspective. The men (and it would have been men) who originally devised the schedule knew how they travelled and they designed around their needs. They didn't deliberately set out to exclude women. They just didn't think about them. They didn't think to consider if women's needs might be different. And so this data gap was a result of not involving women in planning.

Inés Sánchez de Madariaga, an urban-planning professor at Madrid's Technical University, tells me that this is a problem in transport planning more generally. Transport as a profession is 'highly male-dominated', she explains. In Spain, 'the Ministry of Transportation has the fewest women of all the ministries both in political and technical positions. And so they have a bias from their personal experience.'

On the whole, engineers focus mostly on 'mobility related to employment'. Fixed labour times create peak travel hours, and planners need to know the maximum capacity that infrastructure can support. 'So there's a technical reason for planning for peak hours,' Sánchez de Madariaga acknowledges. But needing to plan for peak hours doesn't explain why female travel (which doesn't tend to fit into peak hours, and therefore 'doesn't affect the maximum capacity of systems') gets ignored.

The available research makes bias towards typically male modes of travel clear. The United Nations Commission on the Status of Women found 'a male bias' in transport planning and a failure to address gender 'in system configuration'.[17] A 2014 EU report on Europeans' satisfaction with urban transport describes male travel patterns as 'standard' even as it decries the failure of European public transport systems to adequately serve women.[18] More galling are common planning terms such as 'compulsory mobility', which Sánchez de Madariaga explains is a commonly used umbrella concept for 'all trips made for employment and educational purposes'.[19] As if care trips are not compulsory, but merely expendable 'me time' for dilettantes.

The bias is also clear in government spending priorities. Stephen Bush, the *New Statesman*'s political correspondent, pointed out in a July 2017 article that although the Conservative government has consistently spouted austerity rhetoric, the last two Tory chancellors have made an exception for road-building, on which both have spent lavishly.[20] With living standards falling and Britain already having a fairly serviceable road infrastructure there is a whole host of areas that seem a potentially wiser investment, but somehow, both times, for both men, roads have seemed the obvious choice. Meanwhile, by 2014, 70% of councils had cut bus funding (the most feminised form of transport), with a £19 million cut in 2013 alone, and bus prices had been rising every year.[21]

British politicians are not alone here. A 2007 World Bank report revealed that 73% of World Bank transport funding is for roads and highways, most of them rural or linking up cities.[22] Even where roads are the right investment choice, where the proposed road leads is not a gender-neutral decision. In an illustration of how important it is that development projects are based on sex-disaggregated data, another World Bank report recounted the disagreement over a proposed road in one village in Lesotho. Women wanted the road to be constructed in one direction to 'facilitate their access to the nearest village with basic services'; men wanted it built in the opposite direction 'to enable them to reach the larger town and market more easily on horseback'.[23]

The gender gap in travel data continues with the intentional omission in many transport surveys of shorter pedestrian and other 'non-motorised' trips.[24] These trips, says Sánchez de Madariaga, are 'not considered to be relevant for infrastructure policymaking'. Given women generally walk further and for longer than men (in part because of their care-giving responsibilities; in part because women tend to be poorer), this marginalisation of non-motorised travel inevitably affects them more. Ignoring shorter walking trips also adds to the gap in trip-chaining data, as this kind of travel usually involves at least one journey on foot. In short, the assumption that shorter walking trips are irrelevant to infrastructure policy is little short of an assumption that women are irrelevant to infrastructure policy.

But they aren't. Men tend to travel on their own, but women travel encumbered – by shopping, by buggies, by children or elderly relatives they are caring for.[25] A 2015 survey on travel in London found that women are 'significantly less likely than men to be satisfied with the streets and pavements after their last journey by foot', perhaps reflecting the reality that not only are women more likely to walk than men but also that women are more likely to be pushing prams and therefore be more affected by inadequate walkways.[26] Rough, narrow and cracked pavements littered with ill-placed street furniture combine with narrow and steep steps at numerous transit locations to make travelling around a city with a buggy 'extremely difficult', says Sánchez de Madariaga, who estimates that it can take up to four times as long. 'So what do young women with small kids do?'

Valuing cars over pedestrians is not inevitable. In Vienna 60% of all journeys are made on foot, in no small part because the city takes gender planning seriously. Since the 1990s Vienna's head of gender planning, Eva Kail, has been collecting data on pedestrian travel and has installed the following improvements: improved and signed crossing locations (plus forty additional crossings); retrofitted steps with ramps for prams and bikes; widened 1,000 metres of pavement; and increased pedestrian street lighting.[27]

The mayor of Barcelona, Ada Colau, has shown similar determination to give her city back to pedestrians, creating what are called *superilles* or 'superblocks' – squared-off sections of the city with low speed limits open only to local traffic, with roads where pedestrians have equal priority with cars. Another example of easy changes that can be implemented to accommodate female travel patterns comes via London, where in 2016 the 'hopper fare' was introduced to the bus network.[28] Previously, every time a user boarded a bus they were charged for a new journey, but under the new system users can now make two trips in one hour for the price of one. This change is particularly helpful for women because they were disproportionately penalised by the old charging system. This is not only because of women being more likely to trip-chain, but also because women make up the majority (57%) of London's bus users (partly because it's cheaper, partly because the bus is perceived as more child-friendly), and are more likely to have to transfer (which under the old system counted as a new trip).

The reason women are more likely to have to transfer is because, like most cities around the world, London's public transport system is radial.[29] What this means is that a single 'downtown' area has been identified and the majority of routes lead there. There will be some circular routes, concentrated in the centre. The whole thing looks rather like a spider's web, and it is incredibly useful for commuters, who just want to get in and out of the centre of town. It is, however, less useful for everything else. And this useful/not so useful binary falls rather neatly onto the male/female binary.

But while solutions like London's hopper fare are an improvement, they are by no means standard practice worldwide. In the US, while some cities have abandoned charging for transfers (LA stopped doing this in 2014), others are sticking with it.[30] Chicago for example, still charges for public transport connections.[31] These charges seem particularly egregious in light of a 2016 study which revealed quite how much Chicago's transport system is biased against typical female travel patterns.[32] The study, which compared Uberpool (the car-sharing version of the popular taxi app) with public transport in Chicago, revealed that for trips downtown, the difference in time between Uberpool and public transport was negligible – around six minutes on average. But for trips between neighbourhoods, i.e. the type of travel women are likely to be making for informal work or care-giving responsibilities, Uberpool took twenty-eight minutes to make a trip that took forty-seven minutes on public transport.

Given women's time poverty (women's paid and unpaid work combines into a longer working day than men's), Uberpool might seem attractive.[33] Except it costs around three times more than public transport and women are also cash poor compared to men:

around the world women have less access to household finances than men, while the global gender pay gap currently stands at 37.8% (it varies hugely from country to country, being 18.1% in the UK; 23% in Australia; and 59.6% in Angola).[34]

There is, of course, an issue of resources here, but the problem is, to a certain extent, one of attitude and priorities. Although McKinsey estimates that women's unpaid care work contributes $10 trillion to annual global GDP,[35] trips made for paid work are still valued more than trips made for unpaid care work.[36] But when I ask Sánchez de Madariaga if, in a city like London or Madrid, there is an economic argument for providing transport that caters for women's care responsibilities she replied immediately. 'Absolutely. Women's employment is a really important input to GDP. For every percentage increase in women's employment there is a greater increase in GDP. But for women to work, the city has to support this work.' And one of the key ways to do this is to design transport systems that enable women to do their unpaid work and still get to the office on time.

When it comes to fixed infrastructure like subways and trains, Sánchez de Madariaga explains, there is not much you can easily or cheaply do to address this historical bias. 'You can improve their accessibility,' she says and that's about it. Buses, on the other hand, are flexible and their routes and stops can and should be 'moved and adjusted for need', says Sánchez de Madariaga. This is, in fact, what Ada Colau has done in Barcelona, by introducing a new orthogonal bus route (a grid rather than a spiderweb, which is more useful for trip-chaining). Sánchez de Madariaga also argues that public transport needs to develop 'intermediate services, something between a car and a bus. In Mexico they have something called *perceros*, which are really small, like a mini mini minibus. And they have shared taxis. These have a lot of flexibility, and, I think, could and should be developed to support women's mobility.'

While much of the historical gender data gap in travel planning has arisen simply because the idea that women might have different needs didn't occur to the (mainly) male planners, there is another, less excusable, reason for it, and that is that women are seen as, well, just more *difficult* to measure. 'Women have much more complicated travel patterns,' explains Sánchez de Madariaga, who has designed a survey to measure women's care travel. And on the whole, transport authorities aren't interested in women's 'atypical' travel habits. Anastasia Loukaitou-Sideris, a professor of urban planning at UCLA, tells me that 'oftentimes there is the perception from the part of transit operators that everyone has universal needs. Men, women, everything is the same. And this is completely untrue.' She laughs in exasperation. 'Talking to women riders they bring up a whole slew of different needs that are not being taken care of.'

To make matters worse, transport authorities are compounding the existing gender data gap by failing to separate the data they *do* have by sex. The annual transport statistics report[37] created by the UK government's Department of Transport includes a single statistic (on the gender breakdown of driving-test pass rates – in 2015/16 44% of women passed versus 51% of men), and a link to a page on a government website that hosts a report on gender and walking. The report has nothing to say on the gender breakdown of bus or rail usage, for example, even though this information is vital for planning a transport system that properly serves all its users.

India's public transport agencies also don't separate their data by sex,[38] while a recent EU report bemoaned the paucity of gender-sensitive transport data, explaining that 'this kind of data is not collected on a regular basis in the majority of European countries.'[39] As in the UK, the US's Transport Statistics Annual Report only mentions women twice: once in relation to driving licences and once in relation to walking.[40] Unlike the UK, however, these references are not even presented as usable statistics, just generalised statements.

A more hidden data gap comes courtesy of the way transport agencies around the world present their data. On the whole, all travel for paid work is grouped together into one single category, but care work is subdivided into smaller categories, some of which, like 'shopping', aren't distinguished from leisure. This is failing to sex-disaggregate by proxy. When Sánchez de Madariaga collected care-related travel data in Madrid, she found that the number of trips made for caring purposes almost equalled those made for employment purposes. And when she further refined the data by sex-disaggregating it, she found that care was 'the single and foremost purpose of travel for women, in much the same way as employment is the main purpose of men's travel'. If all travel surveys were to do this, she argues, planners would be forced to take care travel as seriously as employment travel.

If we really want to start designing transport systems that serve women as well as men, it's no good designing transport infrastructure in isolation, cautions Sánchez de Madariaga, because women's mobility is also an issue of overarching planning policy: specifically, the creation of 'mixed use' areas. And mixed-use areas fly in the face of traditional planning norms that, in many countries, legally divide cities into commercial, residential and industrial single-use areas, a practice that is called zoning.

Zoning dates back to antiquity (what was allowed on either side of the city walls, for example), but it wasn't until the Industrial Revolution that we started to see the kind of explicit division of what could be built where that legally separated where you live from where you might work. And, with its oversimplified categories, this kind of zoning has woven a male bias into the fabric of cities around the world.

Zoning laws are based on, and prioritise the needs of, a bread-winning heterosexual married man who goes off to work in the morning, and comes home to the suburbs to relax at night. This is, explains Sánchez de Madariaga, 'the personal reality of most decision-makers in the field', and the idea that the home is mostly a place for leisure 'continues to underpin planning practices throughout the world.'[41]

But if for these decision-makers the home is 'a respite from paid labour' and 'a place for leisure', that is far from its role in most women's lives. Globally women do three times the amount of unpaid care work men do;[42] according to the IMF, this can be further subdivided into twice as much childcare and four times as much housework.[43] In Katebe, a town in central Uganda, the World Bank found that after spending nearly fifteen hours on a combination of housework, childcare, digging, preparing food, collecting fuel and water, women were unsurprisingly left with only around thirty minutes of leisure time per day.[44] By contrast, men, who spent an hour less than women per day digging, negligible amounts of time on housework and childcare, and no time at all on collecting fuel and water, managed to find about four hours per day to spend on leisure. The home may have been a place of leisure for him – but for her? Not so much.

In any case, in most families both parents work, and with women in heterosexual couples being the most likely to have primary caring responsibilities over children and elderly relatives, the legal separation of the home from formal workplaces can make life incredibly difficult. Those who have to accompany children and sick relatives around the peripheries of an urban area poorly served by public transport infrastructure are forgotten. The truth is that most zoning ordinances do not reflect women's lives (or even many men's lives).

The impact of the kind of lazy unthinking that positions the home as a place of leisure can be severe. In 2009, Brazil launched a public housing scheme called Minha Casa, Minha Vida (My House, My Life). The plan was to help those (at the time an estimated 50 million people) living in inadequate housing.[45] It hasn't exactly worked out that way.

The stereotypical image of Brazil's *favelas* is one of substandard slums, of crime-ridden areas of poverty and lawlessness, where cowed residents live in fear of prowling gangs. There is a grain of truth to this stereotype, but for many *favela* residents, the reality is very different, and the homes they live in are simply the community-built social housing the state has failed to provide. They have grown in response to need, and are generally located in convenient locations, for both work and transport.

The same cannot be said for the Minha Casa, Minha Vida (MCMV) complexes, which have mostly been built on the far edges of the West Zone, an area which in 2010 was described by Antônio Augusto Veríssimo, director of Rio's housing ministry, as a '*região dormitório*', a dormant region, because of its lack of jobs.[46] In fact, Veríssimo discouraged the building of public housing in this area, for fear of creating '*mais guetos de pobreza*' – more ghettos of poverty. Research from the London School of Economics has also found that the majority of those who have been resettled have been moved much further from their original homes than the 7 km distance allowed under municipal law.[47]

Luisa, forty-two, used to live in a *favela* in Rio's wealthy South Zone, where, along with the Central and North Zones, the majority of jobs in Rio are to be found. 'I walked out of my door and was practically already at work,' she told a researcher for the Heinrich Böll Foundation.[48] 'There was transportation going everywhere. I didn't have to walk for miles just to get to a bus stop.' She now lives in an MCMV condo in Campo Grande, in Rio's underdeveloped West Zone, more than 50 km away from her old home.

With no jobs in the immediate vicinity, residents must travel up to three hours to the North and Central Zones using a transport infrastructure that can be described as limited at best. Over 60% of the new housing units are a thirty-minute walk from the nearest train or metro.[49] And the failure to provide adequate public transport for those relocated from the centre to the outskirts of Rio impacts on women in particular because Rio follows the global trend of men dominating car ownership: 71% of cars are owned by men, and men are twice as likely as women to travel using individual vehicles.[50]

It also particularly impacts on women because of their unpaid care-work responsibilities. Melissa Fernández Arrigoitia, a researcher at LSE, told me about the panic of a woman she interviewed who had just been told that she was being moved to an MCMV complex. Pregnant and already a mother of two, she was only able to work because she could rely on her mother for childcare. Being moved 70 km away from her mother and her workplace would make keeping her job impossible. And in the new MCMV complexes what little childcare provision exists has 'not been renovated or expanded to attend to the new residents'.[51]

The failure to provide childcare is exacerbated by the design of the government's new complexes. The apartments themselves have been designed for traditional nuclear families – but the nuclear family is by no means the standard family unit in a *favela*. 'It's very rare that you go into a home in a *favela* and there aren't three generations living there,' says Dr Theresa Williamson, a Rio-based urban planning expert, adding that she's 'never seen an elderly person living on their own in a *favela*'. Similarly, the majority of the households Arrigoitia interviewed were single mothers, often with both children and an older parent living with them. But the standardised design of these 'super tiny' housing units 'didn't respond at all to the potential variety of families', and a side effect is that the childcare solution that intergenerational *favela* living often provides has been excluded from the new complexes by design.

As for public space in the MCMV complexes, this is more or less limited to 'huge car parks', despite the fact that very few people have cars, and 'horribly maintained playgrounds' with equipment that is so cheap it is destroyed within a couple of months (and not replaced). The complexes seem designed with privacy rather than community in mind. For the families used to the intimacy of the *favela* where, explains Williamson, 'your kid doesn't necessarily even need childcare after a certain age, because everybody is always watching them', this often translates into isolation and fear of crime. The upshot is that 'kids aren't outside as much, they stay in their apartments'. And 'suddenly women do need to be watching their kids in a way they didn't used to in the *favela*'. Suddenly they need childcare. And they don't have any.

This isn't even an issue of resources. It's an issue of priorities. Brazil spent millions on public transport infrastructure in the run-up to the 2014 World Cup and 2016 Olympics. The money was there, it was just being spent elsewhere. LSE Cities research found that the new Bus Rapid Transit corridors tended to privilege areas where Olympic facilities were located, leaving 'the problem of collective transport between the poorer resettlements and downtown [...] unattended'.[52] Furthermore, according to residents, government relocation priorities seemed to be less about helping those who needed better housing and more about making way for the upcoming World Cup and Olympics infrastructure development.

And so the women pay. Cristine Santos lost her job in a market in Nova Iguaçu after she was moved to the Vivenda Das Patativas complex in Campo Grande. 'I had to take three buses,' she explained.[53] Another woman was so exhausted from her daily commute of up to six hours, that she had a near-fatal car accident.[54] With few other options open to them women have taken to setting up shop in their new homes, selling drinks, preparing lunch plates, cutting people's hair. But they have to do it in the knowledge that it could get them evicted, because in doing so they are flouting zoning regulations. Turning your home into your workplace is an option in *favela* living because there are no zoning regulations in place: the whole area is already technically illegal. This is not the case with the government's public housing, where, being a residential zone, running a business from your home is strictly forbidden.

So, to sum up, the Brazilian government moved women away from the formal workplace (and indeed the informal workplace: women dominate Brazil's 7.2 million domestic workers) and provided them with inadequate public transport and no childcare.[55] In so doing, they practically forced women to turn their homes into their workplace, by making this the one option that is realistically open to them. And they've made it illegal.

Public housing doesn't have to be this way: but the alternative does require thought. When Vienna's public officials decided to build a new housing complex in 1993, they first defined 'the needs of the people using the space' and then looked for technical solutions to meet those needs, explains Eva Kail.[56] What this meant was collecting data, specifically sex-disaggregated data, because the 'people' this housing was intended to serve were women.

Surveys compiled at the time by Austria's national statistics agency revealed that women spent more time per day than men on household chores and childcare.[57] (According to the latest World Economic Forum figures Austrian women spend double the time men spend on unpaid work, and more time overall on paid and unpaid work combined.)[58] And so, explains Kail, officials designed the housing complex Frauen-Werk-Stadt I (Women-Work-City I – there has since been a II and a III) to cater for women's caring needs.

First came the location, which, Kail says, was carefully chosen to make it easier for women to carry out their caring responsibilities. The complex is right next to a tram stop, has a kindergarten on-site and is close to schools, meaning children can travel on their own from an early age (Sánchez de Madariaga tells me that one of the biggest time drains for women is 'escorting kids to school, to doctors, and to extracurricular activities'). A doctor's practice, a pharmacy and commercial space for other shops are all included within the complex, and there is a large supermarket nearby. It is the ultimate in mixed-use design.

The design of FWS I is, in fact, rather like a purpose-built *favela*. It prioritises community and shared space. Interconnected buildings with a maximum of four units per floor stand around a series of shared courtyards (complete with grassy areas and children's play spaces) which are visible from any unit in the project. Meanwhile, transparent stairwells visible to the outdoors, high levels of lighting in public spaces, and well-lit car parking accessible only via flats, were all designed to promote a sense of safety.[59] Another housing complex in Vienna (Autofreie Mustersiedlung) dispensed with parking spaces altogether, bypassing the zoning rule that specifies one car parking space per new apartment.[60] They instead spent the money on communal rooms and additional play areas. The complex was not specifically aimed at women, but given women are less likely to drive and more likely to care for children than men, the outcome is nevertheless one that caters to women's housing and care needs.

Care work is also built into the interior of the open-floor plan FWS I flats. The kitchen is at the heart of each flat, its visible lines of sight to the rest of the home mirroring the outer courtyard design. This not only enables women to keep an eye on children while working in the kitchen, it also places housework at the heart of the house: a subtle challenge to the idea that housework is solely a woman's responsibility. Compare this to the tendency a local official in Philadelphia revealed she had to repeatedly check in developers of putting kitchens up on a third floor with no elevator: 'Do you want to carry your groceries and strollers up to the third floor?' she points out.[61]

CHAPTER 2

Gender Neutral With Urinals

In April 2017 veteran BBC journalist Samira Ahmed wanted to use a toilet. She was at a screening of *I Am Not Your Negro* at London's famous Barbican arts centre, and it was the interval. Any woman who has ever been to the theatre knows what this means: a rush as soon as the lights go up to try to beat the inevitable queue that will soon be snaking it way across the foyer floor.

Women are used to queueing when they go out. It's frustrating and puts a dampener on their evening. No nice interval chit-chat about the show with friends over a drink, just dull, tedious lining up, occasionally leavened by the knowing eye rolls they share with their fellow waiting women.

But this evening was different. This evening, the queue was worse than usual. Far worse. Because in an almost comically blatant display of not having thought about women at all, the Barbican had turned both the male and female toilets gender neutral simply by replacing the 'men' and 'women' signage with 'gender neutral with urinals' and 'gender neutral with cubicles'. The obvious happened. Only men were using the supposedly 'gender neutral with urinals' and everyone was using the 'gender neutral with cubicles'.

Rather than rendering the toilets actually gender neutral by this move, they had simply increased the provision for men: women are generally not able to use urinals, while men are of course able to use both urinals and cubicles. There were also no sanitary bins in the 'gender neutral with urinals' toilets. 'Ah the irony of having to explain discrimination having just been to see *I Am Not Your Negro* IN YOUR CINEMA', Ahmed tweeted, suggesting that the solution would be to 'turn the gents into gender-neutral loos. There's NEVER such a queue there & you know it.'[1]

Although this truism seems to have passed the Barbican's heavily male-dominated management team by, it is true that the perennial queueing problem is one that men do tend to know about – given it so often spills out of the main bathroom door, it's hard for even the most oblivious man to miss.[2] But fewer people – men or women – know exactly why it happens. There is a tendency (as ever) to blame the women rather than male-biased design. But male-biased design is in fact exactly what the problem is here.

On the face of it, it may seem fair and equitable to accord male and female public toilets the same amount of floor space – and historically, this is the way it has been done. 50/50 division of floor space has even been formalised in plumbing codes. However, if a male toilet has both cubicles and urinals, the number of people who can relieve themselves at once is far higher per square foot of floor space in the male bathroom than in the female bathroom. Suddenly equal floor space isn't so equal.

But even if male and female toilets had an equal number of stalls, the issue wouldn't be resolved, because women take up to 2.3 times as long as men to use the toilet.[3] Women make up the majority of the elderly and disabled, two groups that will tend to need more time in the toilet. Women are also more likely to be accompanied by children, as well as disabled and older people.[4] Then there's the 20–25% of women of childbearing age who may be on their period at any one time, and therefore needing to change a tampon or a sanitary pad.

Women may also in any case require more trips to the bathroom than men: pregnancy significantly reduces bladder capacity, and women are eight times more likely to suffer from urinary-tract infections than men which again increases the frequency with which a toilet visit is needed.[5] In the face of all these anatomical differences, it would surely take a formal (rather than substantive) equality dogmatist to continue to argue that equal floor space between men and women is fair.

It gets a lot worse than supposedly equal provision being in fact male-biased. A third of the world's population lack adequate toilet provision at all.[6] According to the UN, one in three women lack access to safe toilets,[7] and WaterAid reports that girls and women collectively spend 97 billion hours a year finding a safe place to relieve themselves.[8] The lack of adequate toilet provision is a public health problem for both sexes (for example, in India, where 60% of the population does not have access to a toilet,[9] 90% of surface water is contaminated[10]), but the problem is particularly acute for women, in no small part because of the attitude that men can 'go anywhere',[11] while for women to be seen urinating is shameful. Women get up before dawn and then wait for hours until dusk to go out again in search of a relatively private place to urinate or defecate.[12] And this isn't just a problem in poor countries: Human Rights Watch spoke to young girls working in tobacco fields in America and found that they would 'refrain from relieving themselves at all during the day – aided by avoiding drinking liquids, which increased their risk of dehydration and heat illness'.[13]

This affects women's paid labour: women make up 91% of the 86% of Indians who work in the informal economy. Many of these women work as market vendors, and no public toilets means they have nowhere to go during the workday.[14] In Afghanistan, female police officers go to the toilets in pairs, because their changing and toilet facilities (described by an international advisor to Human Rights Watch as 'a site of harassment') often have peepholes or doors which don't lock. The lack of safe toilet provision in fact often prevents women from joining the force at all, and this in turn has had a significant impact on how the police respond to crimes against women and girls.[15]

Despite women's arguably greater need for public sanitary facilities, however, men are often the ones who are better provided for. More than half of Mumbai's 5 million women do not have an indoor toilet and there are no free public toilets for women. Meanwhile, free urinals for men run into the thousands.[16] A typical Mumbai slum might have six bathrooms for 8,000 women,[17] and government figures from 2014 revealed that the city as a whole has '3,536 public restrooms that women share with men, but not a single women's-only facility – not even in some police stations and courts'.[18]

A 2015 survey found that 12.5% of women in Mumbai's slums defecate in the open at night: they 'prefer to take this risk to walking 58 metres, the average distance of the community toilet from their homes'.[19] But defecating in the open isn't really much safer for women: there is a real danger of sexual assault from men who lurk near and on the routes to areas which are known to be used by women when they need to relieve themselves.[20] The level of violation ranges from voyeurism (including being masturbated at) to rape – and in extreme cases, to murder.

Accurate data on the level of sexual harassment and assault faced by women as they seek to engage in what should be a mundane activity is hard to come by, in no small degree because of the shame surrounding the issue. Few women are willing to talk about something they may well be blamed for 'encouraging'.[21] But what data does exist makes it clear that a failure to provide adequate sanitation is a feminist issue.

A 2016 study found that Indian women who use fields to relieve themselves are twice as likely to face non-partner sexual violence as women with a household toilet.[22] Following the 2014 murder of two girls aged twelve and fourteen in Uttar Pradesh,[23] there was a brief flurry of national focus on the lack of adequate toilet provision for women, and in December 2014, Bombay's high court ordered all municipal corporations to provide safe and clean toilets for women near main roads.[24] Ninety-six potential sites were identified and Bombay's local government promised 50 million rupees (around £550,000) to build new toilets. But a year later, reported online women's rights magazine *Broadly*, not a single brick had been laid.[25] The fund allocation lapsed in 2016.[26]

Local governments that fail to provide public toilets may believe that they are cutting costs, but a 2015 Yale study suggests that this is a false economy. The study authors developed a mathematical model linking the 'risk of sexual assault to the number of sanitation facilities and the time a woman must spend walking to a toilet', and calculated the tangible costs (lost earnings, medical, court and prison expenses) and intangible costs (pain and suffering, risk of homicide) of sexual assault versus the cost of installing and maintaining toilets.

They applied their model to Khayelitsha, a township in South Africa, which has an estimated 5,600 toilets for a population of 2.4 million, resulting, the authors claimed, in 635 sexual assaults at a cost of $40 million each year. Increasing the number of toilets to 11,300, at a direct cost of $12 million, would almost half the average distance to a toilet and result in a 30% decrease in sexual assault. According to the mathematical model, the reduced social and policing costs more than offset the additional cost of providing toilets, leaving the township $5 million better off. These figures, they added, were conservative, since their costings had not included 'the many additional health benefits of improving sanitation in resource-constrained urban areas'.[27]

And there *are* many additional health benefits, particularly for women. Women get bladder and urinary-tract infections from holding in their urine; others suffer from dehydration or chronic constipation.[28] Women who defecate outdoors are at risk of a range of infections and diseases, including pelvic inflammatory disease, worm infections, hepatitis, diarrhoea, cholera, polio and waterborne diseases. Some of these diseases kill millions of people (particularly women and children) every year in India alone.[29]

Health problems arising from a lack of public sanitary provision are not restricted to low-income countries. Canadian and British studies have revealed that referrals for urinary-tract infections, problems with distended bladders, and a range of other urogynaecological problems have increased proportionately to toilet closure; similarly, research shows that the chances of streptococcal toxic shock syndrome from sanitary protection are increased 'if there are no toilets available to change tampons during menstruation'.[30] And, increasingly, there *isn't* a toilet available. A 2007 study revealed that public-toilet closure in the US has been a trend for over half a century.[31] In the UK, 50% of public toilets were closed between 1995 and 2013 – or, as in the public toilet closest to where I live in London, converted into the proverbial hipster bar.[32]

Urban planning that fails to account for women's risk of being sexually assaulted is a clear violation of women's equal right to public spaces – and inadequate sanitary provision is only one of the many ways planners exclude women with this kind of gender-insensitive design.

Women are often scared in public spaces. In fact, they are around twice as likely to be scared as men. And, rather unusually, we have the data to prove it. 'Crime surveys and empirical studies from different parts of the world show that a majority of women are fearful of the potential violence against them when in public spaces,' explains urban-planning professor Anastasia Loukaitou-Sideris. Analyses of crime data from the US and Sweden both show that women and men respond to similar environmental conditions differently, with women tending to be 'more sensitive than men to signs of danger and social disorder, graffiti, and unkempt and abandoned buildings'.

A UK Department for Transport study highlighted the stark difference between male and female perceptions of danger, finding that 62% of women are scared walking in multistorey car parks, 60% are scared waiting on train platforms, 49% are scared waiting at the bus stop, and 59% are scared walking home from a bus stop or station. The figures for men are 31%, 25%, 20% and 25%, respectively.[33] Fear of crime is particularly high among low-income women, partly because they tend to live in areas with higher crime rates, but also because they are likely to be working odd hours[34] and often come home from work in the dark.[35] Ethnic-minority women tend to experience more fear for the same reasons, as well as having the added danger of (often gendered) racialised violence to contend with.

This fear impacts on women's mobility and their basic right of access to the city.[36] Studies from Finland, Sweden, the United States, Canada, Taiwan and the UK all show that women adjust their behaviour and their travel patterns to accommodate this fear.[37] They avoid specific routes, times and modes of transport. They avoid travelling at night. In one Canadian study exactly half of the women surveyed 'indicated that fear prevents them from using public transportation or parking garages'[38] and studies from around the world find that fear of crime is 'amongst the most important reasons women choose not to use public transport'.[39] If they can afford to, they choose to drive or take a taxi instead.

The trouble is, many of them can't afford to. Most passengers are 'transit captives', meaning that they have no reasonable means other than public transport to get from one place to another.[40] This lack of choice particularly affects low-income women, and those living in the global south – in India, for example, women have limited access[41] to private transport and therefore rely on public transport to a far greater extent than men.[42] These women adopt strategies such as taking a longer roundabout route or only travelling while accompanied. Some women go as far as quitting their jobs – a solution that is not limited to those on low incomes.[43] When I tweeted about women's experiences of harassment on public transport, one man replied to tell me about 'a very intelligent and capable woman' he knows, who 'gave up a really good job in the City and moved out of London because she hated being groped on the Tube'.

Clearly, there is an injustice here. But all too often the blame is put on women themselves for feeling fearful, rather than on planners for designing urban spaces and transit environments that make them feel unsafe. And, as usual, the gender data gap is behind it all. The official statistics show that men are in fact more likely to be victims of crime in public spaces, including public transport. And this paradox, says Loukaitou-Sideris, 'has led to the conclusion that women's fear of crime is irrational and more of a problem than crime itself'. But, she points out, the official statistics do not tell the whole story.

As women navigate public spaces, they are also navigating a slew of threatening sexual behaviours. Before we even get to the more serious offences like being assaulted, women are dealing on a daily basis with behaviours from men that make – and are often calculated to make – them feel uncomfortable. Ranging from catcalling, to being leered at, to the use of 'sexualised slurs [and] requests for someone's name', none of these behaviours is criminal exactly, but they all add up to a feeling of sexual menace.[44] A feeling of being watched. Of being in danger – and in fact these behaviours can easily escalate. Enough women have experienced the sharp shift from 'Smile, love, it might never happen,' to 'Fuck you bitch why are you ignoring me?' to being followed home and assaulted, to know that an 'innocent' comment from a male stranger can be anything but.

But women don't report these behaviours, because who could they report them to? Until the emergence of groups like 'EverydaySexism' and 'Hollaback', which give women a space in which they can talk about the intimidating-but-just-short-of-criminal behaviours they face in public spaces on a daily basis, public awareness of this behaviour was more or less non-existent. When police in Nottingham started recording misogynistic behaviour (everything from indecent exposure, to groping, to upskirting) as a hate crime (or if the behaviour was not strictly criminal, a hate incident), they found reports shot up – not because men had suddenly got much worse, but because women felt that they would be taken seriously.[45]

The invisibility of the threatening behaviour women face in public is compounded by the reality that men don't do this to women who are accompanied by other men – who are in any case also much less likely to experience this kind of behaviour. A recent Brazilian survey found that two-thirds of women had been victims of sexual harassment and violence while in transit, half of them on public transportation. The proportion among men was 18%.[46] So men who didn't do it and didn't experience it simply didn't know it was going on. And they all too often dismissed women who told them about it with an airy 'Well I've never seen it.' Another gender data gap.

And one that is exacerbated by how we collect the data. 'Large-scale data for the prevalence of sexual harassment is lacking', explains a 2017 paper, not only because of under-reporting, but also because it is 'often not included in crime statistics'.[47] Added to this is the problem that sexual harassment 'is often poorly classified', with many studies failing to either 'define harassment or codify harassment types'. In 2014, the Australia Institute found that 87% of the women surveyed had experienced verbal or physical street harassment, but data 'concerning the extent or form of incidences were not collected'.

The apparent mismatch between women's fear and the level of violence the official statistics say they experience is not just about the general stew of menace women are navigating. Women also aren't reporting the more serious offences. A 2016 survey of sexual harassment in the Washington DC metro found that 77% of those who were harassed never reported, which is around the same level found by Inmujeres, a Mexican government agency that campaigns on violence against women.[48]

The reporting rate is even lower in New York City, with an estimated 96% of sexual harassment and 86% of sexual assaults in the subway system going unreported, while in London, where a fifth of women have reportedly been physically assaulted while using public transport, a 2017 study found that 'around 90% of people who experience unwanted sexual behaviour would not report it'.[49] An NGO survey of female metro users in Baku, Azerbaijan found that none of the women who said they had been sexually harassed reported it to the appropriate authority.[50]

Clearly then, official police data is not showing the full picture. But although there is a lack of global data on 'the exact nature, location and time' of sexual crimes against women in public spaces, a growing body of research shows that women are in fact not being irrational.[51]

From Rio to Los Angeles men have raped women and girls on buses while drivers carry blithely along their routes.[52] 'The truth is that every time I leave my house, I am scared,' said Victoria Juárez, a thirty-four-year old woman from Mexico where nine in ten women have experienced sexual harassment while using public transport,[53] and female workers report that men hang around in cars 'to kidnap women getting on and off buses'.[54] Travelling to and from work is, they say, the most dangerous part of their day.

A 2016 study found that 90% of French women had been victims of sexual harassment on public transport;[55] in May that year two men were jailed for an attempted gang rape on a Paris train.[56] A 2016 Washington metro survey found that women were three times more likely than men to face harassment on public transport.[57] In April that year[58] a suspect was identified in an indecent exposure incident on the Washington metro; a month later he had escalated to raping a woman at knifepoint on a train.[59] In October 2017 another repeat offender was arrested on the Washington metro: he had targeted the same victim twice.[60]

'The message is unanimous across all articles of this special issue', wrote professor of urban planning Vania Ceccato in her afterword to a 2017 special issue of the academic journal *Crime Prevention and Community Safety*, 'Women's Victimisation and Safety in Transit Environments': 'sexual crime against women in transit (cases of staring, touching, groping, ejaculation, exposing genitalia and full rape) is a highly under-reported offence'.[61]

Women don't report for a variety of reasons. Some of these are societal: stigma, shame, concern that they'll be blamed or disbelieved. And there is little that authorities can do about this. That change has to come from society itself. But many women don't report for more prosaic issues that can be far more easily addressed.

For a start, women often aren't sure exactly 'what counts as sexual harassment and are afraid of the response of authorities'.[62] Assuming they do realise that what has happened is wrong, they often don't know who it is they have to report *to*.[63] Around the world there is a lack of clear information for women on what to do if they are sexually harassed or assaulted on public transport (although most authorities seem to have managed to install clear signage about what to do in the event of spotting a suspicious package). Sometimes, though, the lack of signage is because there really aren't any procedures in place.[64] And this leads to the next problem: the experiences of those women who *do* report.

In 2017 a Danish woman tweeted about what happened when she reported a man who was sexually harassing her on a London bus.[65] After asking her what she expected him to do, the bus driver commented, 'You're a pretty girl, what do you expect?' Her experience echoes that of a twenty-six-year-old woman riding a bus in Delhi: 'It was around 9 p.m. A man standing behind touched me inappropriately. I shouted and caught the guy by his collar. I made the driver stop the bus too. But I was told to get off and solve it myself because other passengers were getting late.'[66]

Fear of being dismissed was why Sarah Hayward, a former local councillor for my borough in London, didn't report. 'I was felt up on a packed Tube train when I was about twenty-two. I can't begin to explain the absolute terror of that feeling. And I just knew that if I said anything, people would think it was just that the Tube was packed.' The irony is, the Tube having been packed may well have been a factor in what happened to her: the data we have suggests that peak travel times coincide with peak sexual harassment times.[67] Hayward tells me that she still tries 'to avoid the Tube in rush hour'.

The lack of reporting procedures for sexual assault is also a problem in the sky. A 2016 *Slate* article told the story of Dana T. who, mid-flight between the US and Germany, woke up to find a hand squeezing her breast hard.[68] It belonged to the man sitting next to her. She told cabin crew who initially tried to make her sit back down. Eventually, they gave her a seat in business class, but although many of the crew were sympathetic, no one seemed to know what to do. When they landed, the man simply got off the plane and went on his way. A similar story emerged in 2017: American Airlines crew refused to move a woman to another seat when it became clear the man next to her was masturbating.[69]

The first step for transit authorities – which have a hugely male-dominated workforce from top to bottom – is to accept that they have a problem.[70] When Loukaitou-Sideris wanted to find out how US transit agencies address women's safety on public transport, she came across a gender data gap. She found only two papers from the 1990s, neither of which looked at the security needs of female passengers and which in any case were redundant given the huge changes that have been made to transport security post-9/11. There was a more recent paper from 2005, but it focused primarily on the response of US transit agencies to the threat of terrorism, 'and did not investigate women's concerns or their specific security needs'.

So Loukaitou-Sideris conducted her own survey. And she encountered some resistance from the male-dominated workforce she surveyed. 'You're assuming that the world is less safe for females,' replied the male chief operating officer of one agency. The male safety and security manager of another insisted that 'Safety and security issues and concerns are non-gender specific.' And in a clear example of the damage the gender data gap does, another (male) safety and security officer refuted the need for gendered planning on the basis that 'Statistical data for our system does not show females have a greater risk.'

Once they have accepted that they have a problem, step two for transport planners is to design evidence-based solutions. Of the 131 transit agencies (more than half of all the large and medium-sized transit operators in the US) that responded to Loukaitou-Sideris's survey, 'only one-third felt that transit agencies should really do something about it', and only three agencies *had* actually done anything about it. Perhaps unsurprisingly given the chronic lack of data and research on women's safety in transport settings, Loukaitou-Sideris also found 'a significant mismatch between the safety and security needs and desires of female passengers and the types and locations of strategies that transit agencies use'.

Most of the agencies she surveyed had security strategies on their buses: 80% had CCTV; 76% had panic alarms; and 73% had public address systems. But the vast majority neither had, nor intended to install, security measures at bus stops. This is in diametric opposition to what women actually want: they are far more likely to feel scared waiting in the dark at a bus stop than they are to feel scared on the bus itself. And in fact, they are right to feel this way: one study found that people were over three times more likely to be a victim of crime at or near a transit stop than on the vehicle itself.[71]

The type of security transport agencies install also matters – and there is also a mismatch here. Transit agencies, possibly for cost reasons, vastly prefer technological solutions to hiring security officers. There is little available data on what impact CCTV has on harassment, but certainly repeated studies have found that women are deeply sceptical of its use, vastly preferring the presence of a conductor or security guard (that is, a preventative solution) as opposed to a blinking light in the corner which may or may not be monitored miles away.[72] Interestingly, men prefer technological solutions to the presence of guards – perhaps because the types of crime they are more likely to experience are less personally violating.[73]

But if paying for a full-time guard is expensive (although arguably worth it if it increases women's use of public transport), there are plenty of cheaper solutions available.[74] Loukaitou-Sideris tells me that 'the city of Portland has a digital timetable in the bus stop so you know when the next bus is going to come', meaning women don't have to wait for ages in the dark, simply because they don't know the next bus is half an hour away. I admit, when I heard this presented as a radical solution I was shocked – in London it's far more unusual to come across a bus stop *without* a digital timetable.

Other evidence-based[75] solutions include transparent bus shelters for better visibility and increased lighting – not just at bus stops and metro stations themselves, but on the route to them.[76] The location of the bus stop is also important: 'sometimes even moving the bus stops a few feet up or down the block if it is in front of a well-used establishment' can make all the difference, says Loukaitou-Sideris. My personal favourite approach is the introduction of request stops in between official stops for women travelling on night buses: although women make up the majority of bus users overall, they are in the minority when it comes to night buses, and while we don't have data on why exactly this disparity exists, given the data we do have it seems reasonable to conclude that feeling unsafe might have something to do with it.[77]

The good news for transport planners is that, other than increased security guard presence and lighting, none of these measures is particularly costly. And research conducted by Loukaitou-Sideris in Los Angeles found that there were specific bus stops that were hotspots for gender-based crime, suggesting that costs could be kept further in check by focusing on problem areas.[78] All each transport authority would need is its own data – and the will to collect it. But that will is lacking. In the US, Loukaitou-Sideris tells me, 'there is no federal incentive' for transit authorities to collect data. 'They aren't legally obligated to collect it and so they don't.' She doesn't buy what she calls their 'excuse' that they don't have the money.

In India (Delhi was ranked the fourth most dangerous public transport system in the world for women in 2014) following what came to be known as the 'Delhi gang rape', women are taking data collection into their own hands.[79] This assault, which hit headlines around the world, began just after 9 p.m. on 16 December 2012 in south Delhi. Twenty-three-year-old physiotherapy student Jyoti Singh and her friend Avanindra Pandey had just finished watching *Life of Pi* at the cinema when they decided to board one of Delhi's many private buses.[80] Their plan was to go home – but they never got there. The two friends were first severely beaten with a rusty iron rod – and then the gang of six men stared to gang rape Singh. The attack (which included shoving the metal rod inside her) lasted nearly an hour, and was so brutal it perforated her colon.[81] Eventually, having exhausted themselves, the six rapists dumped the semi-conscious friends on the roadside, five miles from where they had boarded the bus.[82] Thirteen days later, Singh died from her injuries. The following year, three women set up a crowd-mapping platform called SafeCity.[83] Women can report the location, date and time they were harassed, as well as what happened, 'so that others can view "hot spots" of such incidents on a map'. The data collected so far is revealing: groping is the most common type of harassment – ahead even of catcalls – and it is most likely to happen on public buses (likely because of overcrowding).

Innovative solutions like this are to be welcomed, but they are not a sufficient substitute for data collected and analysed by professional researchers. And this kind of data is severely lacking in all areas of urban planning, not just transport. A 2016 article in the *Guardian* asking why we aren't designing cities 'that work for women, not just men' cautions that the limited number of urban datasets 'that track and trend data on gender make it hard to develop infrastructure programmes that factor in women's needs'.[84] Even when we do start collecting data, there is no guarantee we will continue to do so indefinitely: in 2008 a UK-based database of research on gender and architecture was set up; by 2012 'Gendersite' had closed for lack of funds.[85] And when we don't collect and, crucially, *use* sex-disaggregated data in urban design, we find unintended male bias cropping up in the most surprising of places.

Most women who use a gym will have experienced that moment of psyching herself up to walk into the free weights area, knowing that many of the men who dominate the space will regard her on a range from nuisance to freak. And yes, you *can* technically just walk in, but there's that extra mental hurdle to clear that most men simply don't face, and it takes a particular kind of self-confidence not to be bothered by it at all. Some days, you just won't feel like it. It's the same story in the outdoor gym in my local park; if it's full of men, I often give it a miss, not relishing the inevitable stares and all too clear sense that I don't belong.

The inevitable reaction from some quarters to such complaints is to tell women to stop being delicate flowers – or for feminists to stop painting women as delicate flowers. And of course some women aren't bothered by the leering and macho posturing. But women who do avoid these spaces are not being irrational, because there are plenty of accounts of hostility from men when women venture into supposedly gender-neutral shared exercise spaces.[86] Like transit environments, then, gyms are often a classic example of a male-biased public space masquerading as equal access.

The good news is that this kind of male bias can be designed out and some of the data collection has already been done. In the mid-1990s, research by local officials in Vienna found that from the age of ten, girls' presence in parks and public playgrounds 'decreases significantly'.[87] But rather than simply shrugging their shoulders and deciding that the girls just needed to toughen up, city officials wondered if there was something wrong with the design of parks. And so they planned some pilot projects, and they started to collect data.

What they found was revealing. It turned out that single large open spaces were the problem, because these forced girls to compete with the boys for space. And girls didn't have the confidence to compete with the boys (that's social conditioning for you) so they tended to just let the boys have the space. But when they subdivided the parks into smaller areas, the female drop-off was reversed. They also addressed the parks' sports facilities. Originally these spaces were encased by wire fencing on all sides, with only a single entrance area – around which groups of boys would congregate. And the girls, unwilling to run the gauntlet, simply weren't going in. Enter, stage right, Vienna's very own Leslie Knope, Claudia Prinz-Brandenburg, with a simple proposal: more and wider entrances.[88] And like the grassy spaces, they also subdivided the sports courts. Formal sports like basketball were still provided for, but there was also now space for more informal activities – which girls are more likely to engage in. These were all subtle changes – but they worked. A year later, not only were there more girls in the park, the number of 'informal activities' had increased. And now all new parks in Vienna are designed along the same lines.

The city of Malmö, Sweden, discovered a similar male bias in the way they'd traditionally been planning 'youth' urban regeneration. The usual procedure was to create spaces for skating, climbing and painting grafitti.[89] The trouble was, it wasn't the 'youth' as a whole who were participating in these activities. It was almost exclusively the boys, with girls making up only 10–20% of those who used the city's youth-directed leisure spaces and facilities. And again, rather than shrugging their shoulders and thinking there was something wrong with the girls for not wanting to use such spaces, officials turned instead to data collection.

In 2010, before they began work on their next regeneration project (converting a car park to a leisure area) city officials asked the girls what they wanted.[90] The resulting area is well lit and, like the Viennese parks, split into a range of different-sized spaces on different levels.[91] Since then, Christian Resebo, the official from Malmö's traffic department who was involved in the project, tells me, 'Two more spaces have been developed with the intention of specifically targeting girls and younger women.'

The benefits of this gender-sensitive approach won't just be felt by girls: it may also be felt by the public purse. In the city of Gothenburg in Sweden, around 80 million kronor is distributed every year to sports clubs and associations. Of course, the funding is meant to benefit everyone equally. But when city officials examined the data, they found that it wasn't.[92] The majority of funding was going to organised sports – which are dominated by boys. Grants benefited boys over girls for thirty-six out of forty-four sports. In total, Gothenburg was spending 15 million kronor more on boys' than girls' sports. This didn't just mean that girls' sports were less well funded – sometimes they weren't provided for at all, meaning girls had to pay to do them privately. Or, if they couldn't afford to pay, girls didn't do sports at all.

Most readers will be unsurprised by the report's conclusion that the failure to invest in girls' sport contributed to poorer mental health in girls. More unexpected, perhaps, is the claim that investing in girls' sport could reduce the health cost of fractures due to osteoporosis. Physical exercise increases young people's bone density, reducing the risk of osteoporosis later in life, with research suggesting it is especially important that young girls begin exercising before puberty.

The total cost to Gothenburg of the estimated 1,000 fractures a year resulting from falls (three-quarters of which are suffered by women) is around 150 million kronor. Women account for over 110 million kronor of this. As the report concludes, '[I]f an increase in the city's support for girls' sports of SEK 15 million can lead to a 14 per cent reduction in future fractures due to osteoporosis, the investment will have paid for itself.'

When planners fail to account for gender, public spaces become male spaces by default. The reality is that half the global population has a female body. Half the global population has to deal on a daily basis with the sexualised menace that is visited on that body. The entire global population needs the care that, currently, is mainly carried out, unpaid, by women. These are not niche concerns, and if public spaces are truly to be for everyone, we have to start accounting for the lives of the other half of the world. And, as we've seen, this isn't just a matter of justice: it's also a matter of simple economics.

By accounting for women's care responsibilities in urban planning, we make it easier for women to engage fully in the paid workforce – and as we will see in the next chapter, this is a significant driver of GDP. By accounting for the sexual violence women face and introducing preventative measures – like providing enough single-sex public toilets – we save money in the long run by reducing the significant economic cost of violence against women. When we account for female socialisation in the design of our open spaces and public activities, we again save money in the long run by ensuring women's long-term mental and physical health.

In short, designing the female half of the world out of our public spaces is not a matter of resources. It's a matter of priorities, and, currently, whether unthinkingly or not, we just aren't prioritising women. This is manifestly unjust, and economically illiterate. Women have an equal right to public resources: we must stop excluding them by design.

PART II

The Workplace

CHAPTER 3

The Long Friday

By the end of the day, 24 October 1975 came to be known by Icelandic men as 'the long Friday'.[1] Supermarkets sold out of sausages – 'the favourite ready meal of the time'. Offices were suddenly flooded with children hopped up on the sweets they had been bribed with in an effort to make them behave. Schools, nurseries, fish factories all either shut down or ran at reduced capacity. And the women? Well, the women were having a Day Off.

1975 had been declared by the UN as a Women's Year, and in Iceland women were determined to make it count. A committee was set up with representatives from Iceland's five biggest women's organisations. After some discussion they came up with the idea of a strike. On 24 October, no woman in Iceland would do a lick of work. No paid work, but also no cooking, no cleaning, no child care. Let the men of Iceland see how they coped without the invisible work women did every day to keep the country moving.

Ninety per cent of Icelandic women took part in the strike. Twenty-five thousand women gathered for a rally (the largest of more than twenty to take place throughout the country) in Reykjavík's

Downtown Square – a staggering figure in a country of then only 220,000 people.[2] A year later, in 1976, Iceland passed the Gender Equality Act, which outlawed sex discrimination in workplaces and schools.[3] Five years later, Vigdís Finnbogadóttir beat three men to become the world's first democratically elected female head of state. And today, Iceland has the most gender-equal parliament in the world without a quota system.[4] In 2017 the country topped the World Economic Forum's Global Gender Gap Index for the eighth year running.[5]

Iceland has also been named by *The Economist* as the best country to be a working woman.[6] And while this is of course something to celebrate, there is also reason to take issue with *The Economist*'s phrasing, because if Iceland's strike does anything it is surely to expose the term 'working woman' as a tautology. There is no such thing as a woman who doesn't work. There is only a woman who isn't *paid* for her work.

Globally, 75% of unpaid work is done by women,[7] who spend between three and six hours per day on it compared to men's average of thirty minutes to two hours.[8] This imbalance starts early (girls as young as five do significantly more household chores than their brothers) and increases as they get older. Even in the country with the highest male unpaid working time (Denmark), men still spend less time on unpaid work than women in the country with the lowest female unpaid working time, Norway.[9]

Whenever I raise the issue of the unpaid-work imbalance between men and women, I am invariably faced with the comment, 'But, surely, it's getting better? Surely men are gradually doing more of their share?' And at an individual level, sure, there are men who are doing more. But at a population level? Well, no, not really, because it turns out that the proportion of unpaid work men do is remarkably sticky. An Australian study found that even in wealthier couples who pay for domestic help, the remaining unpaid work is still distributed at the same male to female ratio, with women still doing the majority of what's left.[10] And as women have increasingly joined the paid labour force men have not matched this shift with a comparative increase in their unpaid work: women have simply increased their total work time, with numerous studies over the past twenty years finding that women do the majority of unpaid work irrespective of the proportion of household income they bring in.[11]

Even when men do increase their unpaid work, it isn't by doing the routine housework[12] that forms the majority of the workload,[13] instead creaming off the more enjoyable activities like childcare. On average, 61% of housework is undertaken by women. In India, for example, five out of women's six daily hours of unpaid labour are spent on housework, compared to men's thirteen minutes.[14] It's also rare for men to take on the more personal, messy, emotionally draining aspects of elder care work. In the UK up to 70% of all unpaid dementia carers are women,[15] and female carers are more likely to help with bathing, dressing, using the toilet and managing incontinence.[16] Women are more than twice as likely as men to be providing intensive on-duty care for someone twenty-four hours a day, and to have been caring for someone with dementia for more than five years.[17] Female carers also tend to receive less support than male carers so they end up feeling more isolated and being more likely to suffer from depression – in itself a risk factor for dementia.[18]

Men, meanwhile, have carried on engaging in leisure pursuits – watching TV, playing sports, playing computer games. US men manage to find over an hour more spare time per day to rest than their female counterparts,[19] while in the UK, the Office for National Statistics found that men enjoy five hours more leisure time per week than women.[20] And an Australian study found that what little leisure time women do have is 'more fractured and combined with other tasks' than men's.[21]

The upshot is that around the world, with very few exceptions, women work longer hours than men. Sex-disaggregated data is not available for all countries, but for those where the data exists, the trend is clear. In Korea, women work for thirty-four minutes longer than men per day, in Portugal it's ninety minutes, in China it's forty-four minutes, and in South Africa it's forty-eight minutes.[22] The size of the gap varies from country to country (the World Bank estimates that in Uganda women work an average of fifteen hours every day to men's average of nine hours), but the existence of a gap remains more or less constant.[23]

A 2010 US study on the imbalance between the amount of unpaid work done by male and female scientists found that female scientists do 54% of the cooking, cleaning and laundry in their households, adding more than ten hours to their nearly sixty-hour work week, while men's contribution (28%) adds only half that time.[24] The women in their data set also did 54% of parenting labour in their households, while male scientists did 36%. In India, 66% of women's work time is spent on unpaid labour, while only 12% of men's work is unpaid. In Italy, 61% of women's work is unpaid compared to 23% of men's. In France, 57% of their work is unpaid compared to 38% of men's.

All this extra work is affecting women's health. We have long known that women (in particular women under fifty-five) have worse outcomes than men following heart surgery. But it wasn't until a Canadian study came out in 2016 that researchers were able to isolate women's care burden as one of the factors behind this discrepancy. 'We have noticed that women who have bypass surgery tend to go right back into their caregiver roles, while men were more likely to have someone to look after them,' explained lead researcher Colleen Norris.[25]

This observation may go some way to explaining why a Finnish study[26] found that single women recovered better from heart attacks than married women – particularly when put alongside a University of Michigan study[27] which found that husbands create an extra seven hours of housework a week for women. An Australian study similarly found that housework time is most equal by gender for single men and women; when women start to cohabit, 'their housework time goes up while men's goes down, regardless of their employment status'.[28]

The Economist isn't alone in forgetting about women's unpaid workload in their discussions of 'work'. When business magazines like *Inc* publish think-pieces telling us that 'science' tells us 'you' shouldn't work more than forty hours a week,[29] or when the *Guardian* informs us that 'your job could be killing you' if you work for more than thirty-nine hours per week, they aren't talking to women, because for women there is no 'if'.[30] Women do work well over this amount. Regularly. And it *is* killing them.

It starts with stress. In 2017 the UK's Health and Safety Executive (HSE) released a report on stress in the workplace which revealed that, in every age range, women had higher rates of work-related stress, anxiety and depression than men.[31] Overall, women were 53% more stressed than men, but the difference was particularly dramatic in the age range thirty-five to forty-four: for men the rate was 1,270 cases per 100,000 workers; for women it was nearly double that, at 2,250 cases per 100,000 workers.

The HSE concluded that this disparity was a result of the sectors women work in (stress is more prevalent in public service industries, such as education, health and social care), as well as 'cultural differences in attitudes and beliefs between males and females around the subject of stress'. These may well be part of the reason, but the HSE's analysis is sporting a pretty dramatic gender data gap.

Since 1930 the International Labour Organization (ILO) has stipulated that no one should exceed forty-eight hours a week at work, by which they meant paid work.[32] Beyond this number of hours workers start incurring health costs. But there is a growing consensus that things may be a little bit more complicated than that.

A 2011 analysis of data collected on British civil servants between 1997 and 2004 found that working more than fifty-five hours per week significantly increased women's risk of developing depression and anxiety – but did not have a statistically significant impact on men.[33] Even working forty-one to fifty-five hours seemed to increase the probability of mental health problems in women. This was in line with a 1999 Canadian study[34] and a 2017 analysis[35] of six years of data from the Household Income Labour Dynamics of Australia Survey, both of which found that women had to work far fewer paid hours than men before their mental health started to deteriorate.

But it's not only about mental health. Swedish studies have found that moderate overtime work increases women's hospitalisation and mortality rate, but has a protective effect for men.[36] A 2016 US paper on the impact of long work hours over a thirty-two-year period found a similar gender disparity.[37] Working moderately long hours (forty-one to fifty hours per week) was 'associated with less risk of contracting heart disease, chronic lung disease, or depression' in men. By contrast, such hours for female workers led to consistent and 'alarming increases' in life-threatening diseases, including heart disease and cancer. Women's risk of developing these diseases started to rise when they worked more than forty hours per week. If they worked for an average of sixty hours per week for over thirty years, their risk of developing one of these diseases tripled.

So, what's going on? Is this all proof that women are in fact the weaker sex?

Not exactly. In fact, the Australian study found that although the average man could work substantially longer hours than the average woman before his mental health was negatively impacted, there was one group of workers for whom the gender gap was much narrower. These workers are called the 'unencumbered', that is, workers with little to no care responsibilities. For the unencumbered, both men's and women's work-hour thresholds were much closer to the forty-eight hours stipulated by the ILO. The problem is, women aren't unencumbered. It's just that the work they do is invisible.

When Ryan Gosling thanked his partner Eva Mendes at the 2017 Golden Globes for her unpaid work, acknowledging that without it he would not be on stage accepting an award, he marked himself out as a rare man.[38] Far more usual is the impressively unperceptive man *Guardian* columnist Hadley Freeman wrote about in 2018: '"I have kids and I work full-time," one boss crossly told a friend of mine who asked to have Fridays off. "Yes, and your wife quit her job to look after the kids," my friend couldn't quite bring herself to reply.'[39]

This man simply couldn't see – or perhaps didn't want to see – all the unpaid work that gets done around him. The unpaid work that enables him to have kids *and* easily work full-time in paid employment. It doesn't occur to him that the reason he doesn't need Fridays off is not that he's better than his female co-worker, but rather that, unlike him, she doesn't have a full-time wife at home.

Of course most male bosses in heterosexual relationships won't have a full-time wife at home, because most women can't afford to quit work entirely. Instead, women accommodate their care responsibilities by going part-time. In the UK, 42% of women compared to 11% of men work part-time, and women make up 75% of part-time workers.[40] And part-time work is paid less per hour than full-time work – in part because it's rare that a high-level post is offered as a job-share or with flexible working hours. Women end up working in jobs below their skill level that offer them the flexibility they need[41] – but not the pay they deserve.[42]

In Scotland in 2016 the average hourly gender wage gap was 15% – but this average hid the substantial disparity between full-time and part-time work.[43] For those in full-time work the hourly gap went down to 11%, but the hourly pay gap between men working full-time and women working part-time was 32%. In 2017, median hourly pay for full-time employees across the UK was £14 per hour,[44] compared with £9.12 for part-time employees.[45]

Some call women's segregation into low-paid work a choice. But it's a funny kind of choice when there is no realistic option other than the children not being cared for and the housework not getting done. In any case, fifty year's worth of US census data[46] has proven that when women join an industry in high numbers, that industry attracts lower pay and loses 'prestige',[47] suggesting that low-paid work chooses women rather than the other way around.

This choice-that-isn't-a-choice is making women poor. A recent Organisation for Economic Co-operation and Development (OECD) study found that the gender pay gap in hourly wages is substantially higher in countries where women spend a large amount of time on unpaid care compared to men.[48] In the UK, women make up 61% of those earning below the living wage,[49] and the Institute for Fiscal Studies has found that the gender pay gap widens over the twelve years after a child is born to 33%, as women's careers – and wages – stagnate.[50] The US pay gap between mothers and married fathers is three times higher than the pay gap between men and women without children.[51]

Over time, these pay gaps add up. In Germany a woman who has given birth to one child can expect to earn up to $285,000 less by the time she's forty-five than a woman who has worked full-time without interruption.[52] Data from France, Germany, Sweden and Turkey shows that even after accounting for social transfers that some countries employ to recognise the contribution women make through their unpaid care work, women earn between 31% and 75% less than men over their lifetimes.[53]

This all leaves women facing extreme poverty in their old age, in part because they simply can't afford to save for it. But it's also because when governments are designing pension schemes, they aren't accounting for women's lower lifetime earnings. This isn't exactly a data gap, because the data does mostly exist. But collecting the data is useless unless governments use it. And they don't.

Largely on the advice of international financial institutions such as the World Bank, the last two decades have seen an increasing global shift from social insurance to (often privately managed) individual capital account schemes.[54] The payments a pensioner receives are directly based on their past contributions and the number of years during which the person is expected to collect benefits. This means women are penalised for the following: having to take time out for unpaid care work; early retirement (still a legal requirement in certain countries and professions); and for living longer.

Other policies simply benefit men more than women. These include Australia's recent tax concessions for pension funds (men are likely to have a higher pension pot),[55] and the UK's recent shift to auto-enrolment. As with many pensions around the world, this policy makes the standard error of forgetting to compensate women for the time they have to take out of the paid labour force to attend to their unpaid care load. As a result, women 'miss out on vital contributions to their pension'.[56] More unforgivable is the British system's failure to account for the fact that women are more likely to have several part-time jobs in order to combine their paid and unpaid workloads.[57] In order to qualify for the auto-enrolment pension, a worker must earn at least £10,000 a year. But while many women do earn past this threshold, they earn it from multiple employers – and combined earnings are not counted towards the threshold. This means that '32% or 2.7 million employed women will not earn enough to benefit from auto-enrolment compared to 14% of employed men'.[58]

A counterpoint is provided by Brazil, Bolivia and Botswana, which have achieved close to universal pension coverage and smaller gender gaps 'thanks to the introduction of widely available non-contributory pensions'.[59] Women in Bolivia are credited with one year of pension contributions per child, up to a maximum of three children. As a side benefit (and a more long-term solution to the problem of feminised poverty), pension credits for the main carer have also been found to encourage men to take on more of the unpaid care load.[60] Which raises the question: is women's unpaid work under valued because we don't see it – or is it invisible because we don't value it?

Alongside addressing male bias in pensions, governments must address feminised poverty in old age by introducing policies that enable women to stay in paid work. That starts – but certainly doesn't end – with properly paid maternity leave.

EU countries with comprehensive support for working parents have the highest rates of female employment.[61] Numerous studies world wide have shown that maternity leave has a positive impact on women's participation in the paid labour market.[62] This impact is seen not only in the raw numbers of women employed, but also in the number of hours they work and the income they earn. It has been shown to be particularly beneficial for low-income women.[63]

There is a caveat, however: not all maternity-leave policies are made equal. The length of time and the amount of money on offer matter. If women aren't given enough time off, there is a risk they will leave the paid labour force entirely,[64] or transition to part-time work.[65] When Google noticed that they were losing women who had just given birth at twice the rate of other employees, they increased their maternity leave from three months at partial pay to five months at full pay. The attrition rate dropped 50%.[66]

With the exception of the US, all industrialised countries guarantee workers paid maternity leave,[67] but most countries aren't hitting the sweet spot either in pay or the length of leave allowed. And they certainly aren't hitting them both together. A recent Australian analysis found that the optimum length of paid maternity leave for ensuring women's continued participation in paid labour was between seven months to a year,[68] and there is no country in the world that offers properly paid leave for that length of time.

Twelve countries in the OECD offer full replacement wages, but none of these countries offers more than twenty weeks, with the average being fifteen weeks. Portugal, for instance, one of the countries that offers 100% replacement wages, offers only six weeks of leave. Australia, by contrast offers eighteen weeks of maternity leave – but at 42% of earnings. Ireland offers twenty-six weeks – but at only 34% of earnings. For women in these countries the full length of time they're technically allowed to take off can be, as a result, academic.

British politicians like to boast (particularly in the run-up to the EU referendum) that the UK offers a 'more generous' maternity leave than the fourteen weeks mandated by the EU's 1992 Pregnant Workers Directive.[69] This is technically true, but it doesn't mean that women in the UK get a good deal in comparison to their European counterparts. The average length of paid maternity leave across the EU is twenty-two weeks.[70] This figure hides substantial regional variation in both pay and length. Croatia offers thirty weeks at full pay, compared to the UK's offering of thirty-nine weeks at an average of 30% pay. In fact a 2017 analysis placed the UK twenty-second out of twenty-four European countries on the length of 'decently paid maternity leave' it offered its female workforce (1.4 months).

And now that Britain is leaving the EU, the country is likely to fall even further below its European neighbours. Since 2008, the EU has been trying to extend its maternity-leave ruling to twenty weeks on full pay.[71] This proposal was stuck in stalemate for years, and finally abandoned in 2015 thanks in no small part to the UK and its business lobby, which campaigned strenuously against it.[72] Without the UK, the women of the EU will be free to benefit from this more progressive leave allowance. Meanwhile Martin Callanan (now a Brexit minister) made a speech to the European Parliament in 2012 in which he included the Pregnant Workers Directive in his list of the 'barriers to actually employing people' which 'we could scrap'.[73]

For some women in Britain, no maternity leave at all is already a reality, because the Pregnant Workers Directive doesn't cover female politicians. Women in the national Parliament have access to maternity leave, but there is no provision for them to vote without turning up in person. Technically, women on maternity leave can make use of a system called 'pairing', where one MP is matched up with an MP who would vote in the opposite direction, and neither vote. However, in July 2018 we saw just how inadequate this solution is, when the Conservative MP Brandon Lewis, who was paired with the Lib Dem MP Jo Swinson, mysteriously 'forgot' he was paired when it came to two crucial Brexit votes that the government won by an extremely narrow margin.

But bad as this is, it's even worse in local government. Under Section 85 of the Local Government Act 1972, 'if a councillor does not attend council for six months, they lose their position unless the authority has approved their absence'. You might hope that an approved absence would include maternity leave, but a report commissioned by women's charity the Fawcett Society found that only twelve councils (4%) in England have a formal maternity leave policy, and although some have informal arrangements, three-quarters offer nothing at all.[74] And so, as a result of policies which forget that half the population can and often do give birth, women lose their jobs.

In 2015, councillor Charlene McLean had to stay in hospital for months after she gave birth prematurely. Despite having remained in contact with the council, and having been informed she had normal workers' rights, when she returned to work she was told she would have to stand for re-election because she had been off for six months. Even after what happened to McLean, Newham Council did not change its rules to account for women's bodily realities, instead simply opting to ensure that all expectant mothers received the right information about their lack of rights.[75] The following year Brigid Jones, a Birmingham City councillor, was told that she would have to step down from her role as cabinet member for children's services if she became pregnant.

Things are worse for women in the US, which is one of only four countries in the world that doesn't guarantee at least some paid maternity leave.[76] The Family and Medical Leave Act guarantees twelve weeks of unpaid leave – but, amongst other restrictions you are eligible only if you have worked for a business with at least 50 other employees for the past twelve months.[77] As a result, even *unpaid* leave is only available to 60% of the workforce.[78] There is nothing to prevent the remaining 40% of US women being fired. And of course the number of women who can *afford* to take unpaid leave is lower: one in four American mothers return to work within two weeks of giving birth.

For some US women the gaps are filled in at a state or industry level. In January 2016, President Barack Obama gave federal workers six weeks of paid care leave,[79] while four states (California, Rhode Island, New York and New Jersey, along with Washington DC) now offer paid family leave, funded through employee social insurance.[80] Some women are lucky enough to work at companies that offer maternity leave. But even with these gaps plugged, around 85% of US women have no form of paid leave.[81]

There have been various failed attempts to address this through legislation, a recent one being Trump's proposal in the 2018 federal budget to pay new mothers six weeks of unemployment benefit.[82] This did not pass, but even if it had, the length allowed and the amount paid would not be sufficient to impact on women's participation in the paid labour force. And this is something that the US badly needs, as, in contrast to other industrialised nations, US women's paid labour force participation is actually decreasing – with a 2013 study finding that the lack of family-friendly policies accounts for nearly a third of the discrepancy.[83]

And so the US government continues to attempt to find ways to fix this apparently intractable problem. The latest wheeze, however, provides little more than another example of how gender-blind policy can unwittingly discriminate against women.[84] As I write in 2018, Republicans in Congress are getting excited about the idea of letting people collect social security benefits early to pay for maternity leave – and then delaying their retirement payments to offset the costs. It's easy to see why the idea is attractive: it comes without a cost, at least to the government. But it is far from cost-free to women. The gender pay gap and the time women take off to care for children already results in lower social security benefits for women, a problem this policy will exacerbate.[85] And given women live longer and spend more of their later years in ill health they arguably need more money for retirement, not less.[86] As a result, the main impact of this policy would be to compound the problem of feminised old-age poverty.

US universities provide another example of how gender-blind leave policies can end up discriminating against women. US academics in the tenure-track system have seven years to receive tenure after getting their first academic job or they're fired. This system is biased against women – especially women who want to have children, in part because the years between completing a PhD and receiving tenure (thirty to forty) coincide with the years these women are most likely to try for a baby.[87] The result? Married mothers with young children are 35% less likely than married fathers of young children to get tenure-track jobs,[88] and among tenured faculty 70% of men are married with children compared to 44% of women.[89]

Universities have done little to address this – and even those that have tried, have often done so in gender-blind ways that may end up exacerbating the problem they were trying to solve.[90] In the 1990s and early 2000s, a number of US universities adopted what was intended as a family-friendly policy: parents would receive an extra year per child to earn tenure. But it isn't gender-neutral 'parents' who need this extra year. It is specifically mothers. As the University of Michigan's Alison Davis-Blake drily noted in the *New York Times*, 'giving birth is not a gender-neutral event'.[91] While women may be (variously) throwing up, going to the toilet every five minutes, changing nappies or plugged into their breast pump during this extra year, men get to dedicate more time to their research. So instead of giving a leg up to parents, this policy gave a leg up to men, and at women's expense: an analysis of assistant professors hired at the top fifty US economics departments between 1985 and 2004 found that the policies ultimately led to a 22% decline in women's chances of gaining tenure at their first job. Meanwhile men's chances increased by 19%.[92]

The analysis came in a working paper and the totality of its findings have been challenged[93] – but given what we already know about the disparity between mothers and fathers gaining tenure, and given what the data tells us about who actually does the care work (not to mention the gestating-and-giving-birth-and-breast feeding work) there seems little reason not to make such policies dependent on who is actually carrying the child, and/or who the main carer is. To date, this has not happened.

This is not to say that paternity leave is not important. It certainly is. Beyond the simple matter of fairness (fathers should have the right to be involved in their children's lives), the data we have shows that properly paid paternity leave has a positive impact on female employment. At close to 80% by 2016, Sweden has the highest female employment figures in the EU.[94] It also has one of the highest levels of paternity-leave uptake in the world, with nine out of ten fathers taking an average of three to four months' leave.[95] This compares with a more typical OECD level of one in five fathers taking any parental leave at all – falling to one in fifty in Australia, the Czech Republic and Poland.[96]

This disparity is unsurprising: Sweden has one of the most generous (and, when it was introduced, innovative) paternity-leave policies in the world. Since 1995, Sweden has reserved a month of parental leave (paid at 90% of earnings) exclusively for fathers. This month cannot be transferred to the mother: the father must use this leave or the couple lose it from their overall leave allowance. In 2002, this increased to two months and in 2016 it was further increased to three months.[97]

Prior to the introduction of the 'use it or lose it' leave for fathers, only about 6% of men in Sweden took paternity leave, despite the fact that it had been available for them since 1974. In other words, men didn't take the leave on offer until forced to by the government. This pattern has been repeated in Iceland, where the introduction of a 'daddy quota' doubled the amount of leave taken by men, and in South Korea, where the number of men taking leave rose more than threefold following the introduction in 2007 of a father-specific entitlement.[98] Proving, however, that no good data goes unignored, in 2015 the UK government saw fit to introduce a shared parental leave policy with no allowance reserved exclusively for men. Predictably, the take-up has been 'woefully low', with just one in a hundred men requesting leave in the twelve months after it was introduced.[99]

The introduction of a daddy quota has not been a marked success in Japan, but this is in no small part due to a design that doesn't account for either the gender pay gap or women's bodily reality. While fathers have two months reserved for them out of a possible fourteen months of shared leave, after the first six months of leave, pay decreases from two-thirds of the parent's salary to just half. Given that women need to recover from pregnancy and giving birth, and may be breastfeeding, they are most likely to take leave first, leaving the higher earner (Japanese men earn on average 27% more than Japanese women) to take the biggest salary hit.[100] It is unsurprising, therefore, that only 2% of Japanese men take the months they are entitled to.[101] Japan's extreme work culture likely also plays a part here – in a country where even holidays are frowned on, fathers report being shamed and penalised at work for taking parental leave.

It is worth persevering, however, because the benefits of policies that enshrine in law equal parental responsibility for a child that, after all, two people have created, are long-lasting. Men who take paternity leave tend to be more involved in childcare in the future[102] – perhaps explaining why a 2010 Swedish study found that a mother's future earnings increase by an average of 7% for every month of leave taken by the father.[103]

Evidence-based parental-leave policies won't fix everything, of course, because women's unpaid workload doesn't begin and end with newborn babies, and the traditional workplace is tailored to the life of a mythical unencumbered worker. He – and it implicitly is a he – doesn't need to concern himself with taking care of children and elderly relatives, of cooking, of cleaning, of doctor's appointments, and grocery shopping, and grazed knees, and bullies, and homework, and bath-time and bedtime, and starting it all again tomorrow. His life is simply and easily divided into two parts: work and leisure. But a workplace predicated on the assumption that a worker can come into work every day, at times and locations that are wholly unrelated to the location or opening hours of schools, childcare centres, doctors and grocery stores, simply doesn't work for women. It hasn't been designed to.

Some companies do try to account for the hidden male bias in the traditional workplace and work day. Campbell Soup offers on-site after-school classes and summer programmes for employees' children.[104] Google offers a stipend for takeout meals in the first three months after a baby is born, subsidised childcare, and has included conveniences like dry cleaners on its campus, so employees can do their errands during the workday.[105] Sony Ericsson and Evernote go further, paying for their employees to have their houses cleaned.[106] Workplaces in the US increasingly provide dedicated spaces for new mothers to breast-pump.[107] American Express will even pay for women to ship their breastmilk home if they have to travel for work while they are breastfeeding.[108]

But companies that remember to account for women are exceptions. When Apple announced its US HQ in 2017 as the 'best office building in the world', this state-of-the-art office was slated to include medical and dental treatment, luxury wellness spas – but not a child daycare centre.[109] Best office in the world for men, then?

The truth is that around the world, women continue to be disadvantaged by a working culture that is based on the ideological belief that male needs are universal. The vast majority of American homemakers (97% of whom are women) in a recent poll[110] indicated that they would go back to work if they could work from home (76%) or if the job offered flexible hours (74%) – rather suggesting that while the majority of US companies claim to offer flexible working,[111] the reality is somewhat different. In fact the number of flexible workers in the US fell between 2015 and 2016 and several major US companies are rescinding their remote work policies.[112] In the UK half of employees would like to work flexibly, but only 9.8% of job ads offer flexible working[113] – and women in particular who request it report being penalised.

Companies also still seem to conflate long hours in the office with job effectiveness, routinely and disproportionately rewarding employees who work long hours.[114] This constitutes a bonus for men. Statistician Nate Silver found that in the US, the hourly wage for those working fifty hours or more – 70% of whom are men – has risen twice as fast since 1984 as hourly pay for those working a more typical thirty-five to forty-nine hours per week.[115] And this invisible male bias is exacerbated in certain countries by tax systems that exempt overtime hours from tax[116] – a bonus for being unencumbered[117] that contrasts sharply with the tax relief on domestic services being trialled in Sweden.[118]

The long-hours bias is particularly acute in Japan where it is not unusual for employees to stay in the office past midnight. This is in part because promotion tends to be based on hours worked, as well as the length of time an employee has spent at a company.[119] It also doesn't hurt to take part in 'nomunication', a play on the Japanese word for drinking (*nomu*), and the English word communication.[120] Technically of course women can do all these things, but it's much more difficult for them. Japanese women spend an average of five hours a day on unpaid labour compared to men who spend about an hour: it's clear who will be free to impress the boss by staying in the office till late, followed by back-slapping drinks at a local strip club.[121]

Women's unpaid workload is compounded in Japan by the two-track career options available in most big Japanese firms: career-track and non-career-track. The non-career-track option is mainly administrative, offers few opportunities for advancement, and is known informally as the 'mommy' track – because 'mommies' don't fit into the kind of work-culture that is required for someone on the career-track.[122] Combined with the impact having children has on a woman's chances of promotion (dependent on her ability to demonstrate loyalty through consecutive years worked at a single company), it is unsurprising that 70% of Japanese women stop working for a decade or more after they have their first child, compared to 30% of American women, with many remaining out of the workforce forever.[123] It is also unsurprising that Japan has the sixth-largest gender gap in employment and the third-largest gender pay gap in the OECD.[124]

Long-hours culture is also a problem in academia – and it is exacerbated by career-progression systems designed around typically male life patterns. An EU report on universities in Europe pointed out that age bars on fellowships discriminate against women: women are more likely[125] to have had career breaks meaning that their 'chronological age is older than their "academic" age'.[125] In an article for the *Atlantic*, Nicholas Wolfinger, co-author of *Do Babies Matter: Gender & Family in the Ivory Tower*, suggested that universities should offer part-time tenure track positions.[126] Primary carers can go part-time, while remaining on the tenure track (in effect doubling their probationary period), with the option of going back to full-time when they can. But while some universities do offer this option, it is still rare and comes with all the poverty problems associated with care-induced part-time work elsewhere.

Some women have taken matters into their own hands. In Germany, Nobel Prize-winning developmental biologist Christiane Nüsslein-Volhard set up a foundation when she realised how disadvantaged her female PhD students with children were compared to their male counterparts.[127] These women were 'committed researchers', and their children were in full-time care during the day. But this wasn't enough to level a field so in thrall to long-hours culture: when childcare ended for the day these women were once again encumbered. Meanwhile, their male and childless female colleagues were 'squeezing in extra reading or research'. And so these women, committed researchers though they were, were dropping out.

Nüsslein-Volhard's foundation aims to put a stop to this leaky pipeline. Honourees receive a month stipend that they can spend on 'anything that alleviates their domestic load: house-cleaning services, time-saving appliances like dishwashers or electric dryers, babysitters for nights and weekends when the daycare center is closed or unavailable'. Recipients must be pursuing graduate or postdoctoral work at German universities. And crucially, and unlike the gender-neutral tenure extension for US academics who take parental leave, they must be women.

Ideological male bias doesn't simply arise at a workplace level: it is woven into the laws that govern how employment works. For example: what counts as a work expense. This is not as objective or as gender neutral a decision as you might think. The expenses that a company will allow its employees to claim back will generally correspond to what that country's government has decided counts as a work expense. And this in turn generally corresponds to the kinds of things men will need to claim. Uniforms and tools are in; emergency day care is out.[128]

In the US, what is an allowable work expense is decided by the IRS, which explains that 'Generally you cannot deduct personal, living, or family expenses.'[129] But what counts as a personal expense is debatable – which is where Dawn Bovasso comes in. Bovasso is one of the few female creative directors in US advertising. She is also a single mother. So when her firm announced that it was hosting a directors' dinner, Bovasso had a decision to make: was this dinner worth the $200 it would cost her for a sitter and travel?[130]

Bovasso's male colleagues on the whole had to do no such mental accounting: yes, men can be single parents, but they are a rare beast. In the UK, 90% of single parents are women.[131] In the US the figure is over 80%.[132] In Bovasso's case, her male colleagues were able to just check their calendar and accept or decline. And most of them accepted. In fact not only did they accept, they also booked the hotel next to the restaurant, so they could drink. And unlike her sitter, this cost was claimable on company expenses.

The implicit bias is clear: expense codes are based on the assumption that the employee has a wife at home taking care of the home and the kids. This work doesn't need paying for, because it's women's work, and women don't get paid for it. Bovasso sums it up: 'You can get $30 for takeout if you work late (because your wife isn't there to cook you dinner) or $30 for Scotch if you want to drink your face off, but you can't get $30 for a sitter (because your wife is at home with the kids).' In the event, Bovasso was able to get her company to cover the cost of her childcare – but as she points out, 'these have been exceptions I've had to ask for'. Which is women all over: always the exception, never the default.

And in any case, not all employers will grant these exceptions. The Fawcett Society's 2017 report on local government in England and Wales found that despite regulations dating from 2003 that call for 'all councils to offer an allowance to cover the caring costs that councillors incur when fulfilling their role', in reality, provision is patchy.[133] Some councils don't reimburse caring expenses at all, and most that do only pay a 'contribution'. Rochdale Borough Council's scheme 'pays just £5.06 per hour, and specifically states that it is "a contribution rather than full reimbursement of carers' expenses" – although this important caveat is notably not made for travel expenses'. Adding to the sense that this is a matter of priorities rather than resources, most local-government meetings take place in the evening (when childcare is most likely to be needed), and although it is standard practice in many countries from the US to Sweden for councillors to remotely attend or vote at meetings, current law does not allow for this cheaper alternative.

It is abundantly clear that the culture of paid work as a whole needs a radical overhaul. It needs to take into account that women are not the unencumbered workers the traditional workplace has been designed to suit, and that while men are more likely to fit into this automaton ideal, increasing numbers of them no longer want to. After all, it is simply a fact that none of us, including businesses, could do without the invisible, unpaid work carers do. So it is time to stop penalising them for doing it. Instead, we must start recognising it, valuing it, and designing the paid workplace to account for it.

CHAPTER 4

The Myth of Meritocracy

For most of the twentieth century there were no female musicians in the New York Philharmonic Orchestra. There were a couple of blips in the 1950s and 60s, when a woman or two was hired, but those aside, the proportion of women sat stubbornly at zero. But then all of a sudden, something changed: from the 1970s onwards, the numbers of female players started to go up. And up.

Turnover in orchestras is extremely low. The composition of an orchestra is fairly static (at around one hundred players), and when you're hired, it's often for life; it's rare that a musician is fired. So there was something remarkable going on when the proportion of women in this orchestra grew from a statistical 0% to 10% in a decade.

That something was blind auditions.[1] Instituted in the early 1970s following a lawsuit, blind auditions are what they sound like: the hiring committee can't see who is playing in the audition, because there is a screen between them and the player.[2] The screens had an immediate impact. By the early 1980s, women began to make up 50% of the share of new hires. Today, the proportion of female musicians in the New York Philharmonic stands at over 45%.[3]

The simple step of installing a screen turned the audition process for the New York Philharmonic into a meritocracy. But in this, it is an outlier: for the vast majority of hiring decisions around the world, meritocracy is an insidious myth. It is a myth that provides cover to institutional white male bias. And, dishearteningly, it is a myth that proves remarkably resistant to all the evidence, going back decades, that shows it up as the fantasy it most certainly is. If we want to kill this myth off, we're clearly going to have to do more than just collect data.

The fact that meritocracy is a myth is not a popular one. Around the industrialised world, people believe that not only is meritocracy the way things *should* work, it's the way things *do* work.[4] Despite evidence suggesting that, if anything, the US is *less* meritocratic than other industrialised countries,[5]Americans in particular hold on to meritocracy as an article of faith, and employment and promotion strategies over the past few decades have increasingly been designed as if meritocracy is a reality. A survey of US firms found that 95% used performance evaluations in 2002 (compared to 45% in 1971) and 90% had a merit-based pay plan in place.[6]

The problem is, there is little evidence that these approaches actually work. In fact, there is strong evidence that they don't. An analysis of 248 performance reviews collected from a variety of US-based tech companies found that women receive negative personality criticism that men simply don't.[7] Women are told to watch their tone, to step back. They are called bossy, abrasive, strident, aggressive, emotional and irrational. Out of all these words, only aggressive appeared in men's reviews at all – 'twice with an exhortation to be more of it'. More damningly, several studies of performance-related bonuses or salary increases have found that white men are rewarded at a higher rate than equally performing women and ethnic minorities, with one study of a financial corporation uncovering a 25% difference in performance-based bonuses between women and men in the same job.[8]

The myth of meritocracy achieves its apotheosis in America's tech industry. According to a 2016 survey, the number one concern of tech start-up founders was 'hiring good people', while having a diverse workforce ranked seventh on the list of ten business priorities.[9] One in four founders said they weren't interested in diversity or work-life balance at all. Which, taken together, points to a belief that if you want to find 'the best people', addressing structural bias is unnecessary. A belief in meritocracy is all you need.

Actually, a belief in meritocracy *may* be all you need – to introduce bias, that is. Studies have shown that a belief in your own personal objectivity, or a belief that you are not sexist, makes you less objective and more likely to behave in a sexist way.[10] Men (women were not found to exhibit this bias) who believe that they are objective in hiring decisions are more likely to hire a male applicant than an identically described female applicant. And in organisations which are explicitly presented as meritocratic, managers favour male employees over equally qualified female employees.

Tech's love affair with the myth of meritocracy is ironic for an industry so in thrall to the potential of Big Data, because this is a rare case where the data actually exists. But if in Silicon Valley meritocracy is a religion, its God is a white male Harvard dropout. And so are most of his disciples: women make up only a quarter of the tech industry's employees and 11% of its executives.[11] This is despite women earning more than half of all undergraduate degrees in the US, half of all undergraduate degrees in chemistry, and almost half in maths.[12]

More than 40% of women leave tech companies after ten years compared to 17% of men.[13] A report by the Center for Talent Innovation found that women didn't leave for family reasons or because they didn't enjoy the work.[14] They left because of 'workplace conditions', 'undermining behaviour from managers', and 'a sense of feeling stalled in one's career'. A feature for the *Los Angeles Times* similarly found that women left because they were repeatedly passed up for promotion and had their projects dismissed.[15] Does this sound like a meritocracy? Or does it look more like institutionalised bias?

That the myth of meritocracy survives in the face of such statistics is testament to the power of the male default: in the same way that men picture a man 80% of the time they think of a 'person', it's possible that many men in the tech industry simply don't *notice* how male-dominated it is. But it's also testament to the attractiveness of a myth that tells the people who benefit from it that all their achievements are down to their own personal merit. It is no accident that those who are most likely to believe in the myth of meritocracy are young, upper-class, white Americans.[16]

If white upper-class Americans are most likely to believe in the myth of meritocracy, it should come as no surprise that academia is, like tech, a strong follower of the religion. The upper ranks of academia – particularly those of science, technology, engineering and maths (STEM) – are dominated by white, middle- and upper-class men. It is a perfect Petri dish for the myth of meritocracy to flourish in. Accordingly, a recent study found that male academics – particularly those in STEM – rated fake research claiming that academia had no gender bias higher than real research which showed it did.[17] Also accordingly, gender bias is in fact plentiful – and well documented.

Numerous studies from around the world have found that female students and academics are significantly less likely than comparable male candidates to receive funding, be granted meetings with professors, be offered mentoring, or even to get the job.[18] Where mothers are seen as less competent and often paid less, being a father can work in a man's favour (a gendered bias that is by no means restricted to academia).[19] But despite the abundance of data showing that academia is in fact far from meritocratic, universities continue to proceed as if male and female students, and male and female academics, are operating on a level playing field.

Career progression in academia depends largely on how much you get published in peer-reviewed journals, but getting published is not the same feat for men as it is for women. A number of studies have found that female-authored papers are accepted more often or rated higher under double-blind review (when neither author nor reviewer are identifiable).[20,21] And although the evidence varies on this point, given the abundant male bias that *has* been identified in academia, there seems little reason not to institute this form of blind academic audition. Nevertheless, most journals and conferences carry on without adopting this practice.

Of course, female academics do get published, but that's only half the battle. Citation is often a key metric in determining research impact, which in turn determines career progression, and several studies have found that women are systematically cited less than men.[22] Over the past twenty years, men have self-cited 70% more than women[23] – and women tend to cite other women more than men do,[24] meaning that the publication gap is something of a vicious circle: fewer women getting published leads to a citations gap, which in turn means fewer women progress as they should in their careers, and around again we go. The citations gap is further compounded by male-default thinking: as a result of the widespread academic practice of using initials rather than full names, the gender of academics is often not immediately obvious, leading female academics to be assumed to be male. One analysis found that female scholars are cited as if they are male (by colleagues who have assumed the P stands for Paul rather than Pauline) more than ten times more often than vice versa.[25]

Writing for the *New York Times*, economist Justin Wolfers noted a related male-default habit in journalists routinely referring to the male contributor as the lead author when in fact the lead author was a woman.[26] This lazy product of male-default thinking is inexcusable in a media report, but it's even more unacceptable in academia, and yet here too it proliferates. In economics, joint papers are the norm – and joint papers contain a hidden male bias. Men receive the same level of credit for both solo and joint papers, but, unless they are writing with other female economists, women receive less than half as much credit for co-authored papers as men do. This, a US study contends, explains why, although female economists publish as much as male economists, male economists are twice as likely to receive tenure.[27] Male-default thinking may also be behind the finding that research perceived to have been done by men is associated with 'greater scientific quality':[28] this could be a product of pure sexism, but it could also be a result of the mode of thinking that sees male as universal and female as niche. It would certainly go some way to explaining why women are less likely to appear on course syllabuses.[29]

Of course before a woman gets to face all these hidden hurdles, she must have found the time to do the research in the first place, and that is by no means a given. We've already discussed how women's unpaid workload outside of paid employment impacts on their ability to do research. But their unpaid workload *inside* the workplace doesn't help either. When students have an emotional problem, it is their female professors, not their male professors they turn to.[39] Students are also more likely to request extensions, grade boosts, and rule-bending of female academics.[31] In isolation, a request of this kind isn't likely to take up much time or mental energy – but they add up, and they constitute a cost on female academics' time that male academics mostly aren't even aware of, and that universities don't account for.

Women are also asked to do more undervalued admin work than their male colleagues[32] – and they say yes, because they are penalised for being 'unlikeable' if they say no. (This is a problem across a range of workplaces: women, and in particular ethnic minority women, do the 'housekeeping' – taking notes, getting the coffee, cleaning up after everyone – in the office as well as at home.[33]) Women's ability to publish is also impacted by their being more likely than their male colleagues to get loaded with extra teaching hours,[34] and, like 'honorary' admin posts, teaching is viewed as less important, less serious, less *valuable*, than research. And we run into another vicious circle here: women's teaching load prevents them from publishing enough, which results in more teaching hours, and so on.

The inequity of women being loaded with less valued work is compounded by the system for evaluating this work, because it is itself systematically biased against women. Teaching evaluation forms are widely used in higher education and they represent another example of a situation where we *have* the data, but are simply ignoring it. Decades of research[35] in numerous countries show that teaching evaluation forms are worse than useless at actually evaluating teaching and are in fact 'biased against female instructors by an amount that is large and statistically significant'.[36] They are, however, pretty good at evaluating gender bias. One of these biases is our old friend 'men are the default human', which shows up in objections to female lecturers straying away from a focus on white men. 'I didn't come out of this course with any more information except gender and race struggles, than I came in with,' complained one student who apparently felt that gender and race were not relevant to the topic at hand: US confederation.[37]

Falling into the trap we encountered in the introduction, of not realising that 'people' is as likely to mean 'women' as it is to mean 'men', another student complained that, 'Although Andrea stated on the first day she would teach a peoples [*sic*] perspective it was not illustrated how much was going to be focused on first nation and women's history.' Incidentally, it's worth taking the implication that this lecturer focused almost exclusively on 'first nations and women's history' with a pinch of salt: a friend of mine got a similarly unhappy review from a male student for focusing 'too much' on feminism in her political philosophy lectures. She had spoken about feminism once in ten classes.

Less effective male professors routinely receive higher student evaluations than more effective female teachers. Students believe that male professors hand marking back more quickly – even when that is impossible because it's an online course delivered by a single lecturer, but where half the students are led to believe that the professor is male and half female. Female professors are penalised if they aren't deemed sufficiently warm and accessible. But if they *are* warm and accessible they can be penalised for not appearing authoritative or professional. On the other hand, appearing authoritative and knowledgeable as a woman can result in student disapproval, because this violates gendered expectations.[38] Meanwhile men are rewarded if they are accessible at a level that is simply expected in women and therefore only noticed if it's absent.

An analysis[39] of 14 million reviews on the website RateMyProfessors.com found that female professors are more likely to be 'mean', 'harsh', 'unfair', 'strict' and 'annoying'. And it's getting worse: female instructors have stopped reading their evaluations in droves, 'as student comments have become increasingly aggressive and at times violent'. A female political history lecturer at a Canadian university received the following useful feedback from her student: 'I like how your nipples show through your bra. Thanks.'[40] The lecturer in question now wears 'lightly padded bras' exclusively.

The teaching evaluation study that revealed women are more likely to be 'mean' also found that male professors are more likely to be described as 'brilliant', 'intelligent', 'smart' and a 'genius'. But were these men actually more in possession of raw talent than their female counterparts? Or is it just that these words are not as gender neutral as they appear? Think of a genius. Chances are, you pictured a man. It's OK – we all have these unconscious biases. I pictured Einstein – that famous one of him sticking his tongue out, his hair all over the place. And the reality is that this bias (that I like to call 'brilliance bias') means that male professors are routinely considered more knowledgeable, more objective, more innately talented. And career progression that rests on teaching evaluations completely fails to account for it.

Brilliance bias is in no small part a result of a data gap: we have written so many female geniuses out of history, they just don't come to mind as easily. The result is that when 'brilliance' is considered a requirement for a job, what is really meant is 'a penis'. Several studies have found that the more a field is culturally understood to require 'brilliance' or 'raw talent' to succeed – think philosophy, maths, physics, music composition, computer science – the fewer women there will be studying and working in it.[41] We just don't see women as naturally brilliant. In fact, we seem to see femininity as inversely associated with brilliance: a recent study where participants were shown photos of male and female science faculty at elite US universities also found that appearance had no impact on how likely it was that a man would be judged to be a scientist.[42] When it came to women, however, the more stereotypically feminine they looked, the less likely it was that people would think they were a scientist.

We teach brilliance bias to children from an early age. A recent US study found that when girls start primary school at the age of five, they are as likely as five-year-old boys to think women could be 'really really smart'.[43] But by the time they turn six, something changes. They start doubting their gender. So much so, in fact, that they start limiting themselves: if a game is presented to them as intended for 'children who are really, really smart', five-year-old girls are as likely to want to play it as boys – but six-year-old girls are suddenly uninterested. Schools are teaching little girls that brilliance doesn't belong to them. No wonder that by the time they're filling out university evaluation forms, students are primed to see their female teachers as less qualified.

Schools are also teaching brilliance bias to boys. As we saw in the introduction, following decades of 'draw a scientist' studies where children overwhelmingly drew men, a recent 'draw a scientist' meta-analysis was celebrated across the media as showing that finally we were becoming less sexist.[44] Where in the 1960s only 1% of children drew female scientists, 28% do now. This is of course an improvement, but it is still far off reality. In the UK, women actually outnumber men in a huge range of science degrees: 86% of those studying polymers, 57% of those studying genetics, and 56% of those studying microbiology are female.[45]

And in any case, the results are actually more complicated than the headlines suggest and still provide damning evidence that data gaps in school curriculums are teaching children biases. When children start school they draw roughly equal percentages of male and female scientists, averaged out across boys and girls. By the time children are seven or eight, male scientists significantly outnumber female scientists. By the age of fourteen, children are drawing four times as many male scientists as female scientists. So although more female scientists are being drawn, much of the increase has been in younger children before the education system teaches them data-gap-informed gender biases.

There was also a significant gender difference in the change. Between 1985–2016, the average percentage of female scientists drawn by girls rose from 33% to 58%. The respective figures for boys were 2.4% and 13%. This discrepancy may shed some light on the finding of a 2016 study which found that while female students ranked their peers according to actual ability, male biology students consistently ranked their fellow male students as more intelligent than better-performing female students.[46] Brilliance bias is one hell of a drug. And it doesn't only lead to students mis-evaluating their teachers or each other: there is also evidence that teachers are mis-evaluating their students.

Several studies conducted over the past decade or so show that letters of recommendation are another seemingly gender-neutral part of a hiring process that is in fact anything but.[47] One US study found that female candidates are described with more communal (warm; kind; nurturing) and less active (ambitious; self-confident) language than men. And having communal characteristics included in your letter of recommendation makes it less likely that you will get the job,[48] particularly if you're a woman: while 'team-player' is taken as a leadership quality in men, for women the term 'can make a woman seem like a follower'.[49] Letters of recommendation for women have also been found to emphasise teaching (lower status) over research (higher status);[50] to include more terms that raise doubt (hedges; faint praise);[51] and to be less likely to include standout adjectives like 'remarkable' and 'outstanding'. Women were more often described with 'grindstone' terms like 'hard-working'.

There is a data gap at the heart of universities using teaching evaluations and letters of recommendation as if they are gender neutral in effect as well as in application, although like the meritocracy data gap more broadly, it is not a gap that arises from a lack of data so much as a refusal to engage with it. Despite all the evidence, letters of recommendation and teaching evaluations continue to be heavily weighted and used widely in hiring, promoting and firing, as if they are objective tests of worth.[52] In the UK, student evaluations are set to become even more important, when the Teaching Excellence Framework (TEF) is introduced in 2020. The TEF will be used to determine how much funding a university can receive, and the National Students Survey will be considered 'a key metric of teaching success'. Women can expect to be penalised heavily in this Excellent Teaching new world.

The lack of meritocracy in academia is a problem that should concern all of us if we care about the quality of the research that comes out of the academy, because studies show that female academics are more likely than men to challenge male-default analysis in their work.[53] This means that the more women who are publishing, the faster the gender data gap in research will close. And we should care about the quality of academic research. This is not an esoteric question, relevant only to those who inhabit the ivory towers. The research produced by the academy has a significant impact on government policy, on medical practice, on occupational health legislation. The research produced by the academy has a direct impact on all of our lives. It matters that women are not forgotten here.

Given the evidence that children learn brilliance bias at school, it should be fairly easy to stop teaching them this. And in fact a recent study found that female students perform better in science when the images in their textbooks include female scientists.[54] So to stop teaching girls that brilliance doesn't belong to them, we just need to stop misrepresenting women. Easy.

It's much harder to correct for brilliance bias once it's already been learnt, however, and once children who've been taught it grow up and enter the world of work, they often start perpetuating it themselves. This is bad enough when it comes to human-on-human recruitment, but with the rise of algorithm-driven recruiting the problem is set to get worse, because there is every reason to suspect that this bias is being unwittingly hardwired into the very code to which we're outsourcing our decision-making.

In 1984 American tech journalist Steven Levy published his bestselling book *Hackers: Heroes of the Computer Revolution*. Levy's heroes were all brilliant. They were all single-minded. They were all men. They also didn't get laid much. 'You would hack, and you would live by the Hacker Ethic, and you knew that horribly inefficient and wasteful things like women burned too many cycles, occupied too much memory space,' Levy explained. 'Women, even today, are considered grossly unpredictable,' one of his heroes told him. 'How can a [default male] hacker tolerate such an imperfect being?'

Two paragraphs after having reported such blatant misogyny, Levy nevertheless found himself at a loss to explain why this culture was more or less 'exclusively male'. 'The sad fact was that there never was a star-quality female hacker', he wrote. 'No one knows why.' I don't know, Steve, we can probably take a wild guess.

By failing to make the obvious connection between an openly misogynistic culture and the mysterious lack of women, Levy contributed to the myth of innately talented hackers being implicitly male. And, today, it's hard to think of a profession more in thrall to brilliance bias than computer science. 'Where are the girls that love to program?' asked a high-school teacher who took part in a summer programme for advanced-placement computer-science teachers at Carnegie Mellon; 'I have any number of boys who really really love computers,' he mused.[55] 'Several parents have told me their sons would be on the computer programming all night if they could. I have yet to run into a girl like that.'

This may be true, but as one of his fellow teachers pointed out, failing to exhibit this behaviour doesn't mean that his female students don't love computer science. Recalling her own student experience, she explained how she 'fell in love' with programming when she took her first course in college. But she didn't stay up all night, or even spend a majority of her time programming. 'Staying up all night doing something is a sign of single-mindedness and possibly immaturity as well as love for the subject. The girls may show their love for computers and computer science very differently. If you are looking for this type of obsessive behavior, then you are looking for a typically young, male behavior. While some girls will exhibit it, most won't.'

Beyond its failure to account for female socialisation (girls are penalised for being antisocial in a way boys aren't), the odd thing about framing an aptitude for computer science around typically male behaviour is that coding was originally seen as a woman's game. In fact, women were the original 'computers', doing complex maths problems by hand for the military before the machine that took their name replaced them.[56]

Even after they were replaced by a machine, it took years before they were replaced by men. ENIAC, the world's first fully functional digital computer, was unveiled in 1946, having been programmed by six women.[57] During the 1940s and 50s, women remained the dominant sex in programming,[58] and in 1967 *Cosmopolitan* magazine published 'The Computer Girls', an article encouraging women into programming.[59] 'It's just like planning a dinner,' explained computing pioneer Grace Hopper. 'You have to plan ahead and schedule everything so that it's ready when you need it. Programming requires patience and the ability to handle detail. Women are 'naturals' at computer programming.'

But it was in fact around this time that employers were starting to realise that programming was not the low-skilled clerical job they had once thought. It wasn't like typing or filing. It required advanced problem-solving skills. And, brilliance bias being more powerful than objective reality (given women were already doing the programming, they clearly had these skills) industry leaders started training men. And then they developed hiring tools that seemed objective, but were actually covertly biased against women. Rather like the teaching evaluations in use in universities today, these tests have been criticised as telling employers 'less about an applicant's suitability for the job than his or her possession of frequently stereotyped characteristics'.[60] It's hard to know whether these hiring tools were developed as a result of a gender data gap (not realising that the characteristics they were looking for were male-biased) or a result of direct discrimination, but what is undeniable is that they *were* biased towards men.

Multiple-choice aptitude tests which required 'little nuance or context-specific problem solving' focused instead on the kind of mathematical trivia that even then industry leaders were seeing as increasingly irrelevant to programming. What they were mainly good at testing was the type of maths skills men were, at the time, more likely to have studied at school. They also were quite good at testing how well networked an applicant was: the answers were frequently available through all-male networks like college fraternities and Elks lodges (a US-based fraternal order).[61]

Personality profiles formalised the programmer stereotype nodded to by the computer-science teacher at the Carnegie Mellon programme: the geeky loner with poor social and hygiene skills. A widely quoted 1967 psychological paper had identified a 'disinterest in people' and a dislike of 'activities involving close personal interaction' as a 'striking characteristic of programmers'.[62] As a result, companies sought these people out, they became the top programmers of their generation, and the psychological profile became a self-fulfilling prophecy.

This being the case, it should not surprise us to find this kind of hidden bias enjoying a resurgence today courtesy of the secretive algorithms that have become increasingly involved in the hiring process. Writing for the *Guardian*, Cathy O'Neil, the American data scientist and author of *Weapons of Math Destruction*, explains how online tech-hiring platform Gild (which has now been bought and brought in-house by investment firm Citadel[63]) enables employers to go well beyond a job applicant's CV, by combing through their 'social data'.[64] That is, the trace they leave behind them online. This data is used to rank candidates by 'social capital' which basically refers to how integral a programmer is to the digital community. This can be measured through how much time they spend sharing and developing code on development platforms like GitHub or Stack Overflow. But the mountains of data Gild sifts through also reveal other patterns.

For example, according to Gild's data, frequenting a particular Japanese manga site is a 'solid predictor of strong coding'.[65] Programmers who visit this site therefore receive higher scores. Which all sounds very exciting, but as O'Neil points out, awarding marks for this rings immediate alarm bells for anyone who cares about diversity. Women, who as we have seen do 75% of the world's unpaid care work, may not have the spare leisure time to spend hours chatting about manga online. O'Neil also points out that 'if, like most of techdom, that manga site is dominated by males and has a sexist tone, a good number of the women in the industry will probably avoid it'. In short, Gild seems to be something like the algorithm form of the male computer-science teacher from the Carnegie programme.

Gild undoubtedly did not intend to create an algorithm that discriminated against women. They were intending to *remove* human biases. But if you aren't aware of how those biases operate, if you aren't collecting data and taking a little time to produce evidence-based processes, you will continue to blindly perpetuate old injustices. And so by not considering the ways in which women's lives differ from men's, both on and offline, Gild's coders inadvertently created an algorithm with a hidden bias against women.

But that's not even the most troubling bit. The most troubling bit is that we have no idea how bad the problem actually is. Most algorithms of this kind are kept secret and protected as proprietary code. This means that we don't know how these decisions are being made and what biases they are hiding. The only reason we know about this potential bias in Gild's algorithm is because one of its creators happened to tell us. This, therefore, is a double gender data gap: first in the knowledge of the coders designing the algorithm, and second, in the knowledge of society at large, about just how discriminatory these AIs are.

Employment procedures that are unwittingly biased towards men are an issue in promotion as well as hiring. A classic example comes from Google, where women weren't nominating themselves for promotion at the same rate as men. This is unsurprising: women are conditioned to be modest, and are penalised when they step outside of this prescribed gender norm.[66] But Google was surprised. And, to do them credit, they set about trying to fix it. Unfortunately the *way* they went about fixing it was quintessential male-default thinking.

It's not clear whether Google didn't have or didn't care about the data on the cultural expectations that are imposed on women, but either way, their solution was not to fix the male-biased system: it was to fix the women. Senior women at Google started hosting workshops 'to encourage women to nominate themselves', Laszlo Bock, head of people operations, told the *New York Times* in 2012.[67] In other words, they held workshops to encourage women to be more like men. But why should we accept that the way men do things, the way men see themselves, is the correct way? Recent research has emerged showing that while women tend to assess their intelligence accurately, men of average intelligence think they are more intelligent than two-thirds of people.[68] This being the case, perhaps it wasn't that women's rates of putting themselves up for promotion were too low. Perhaps it was that men's were too high.

Bock claimed Google's workshops as a success (he told the *New York Times* that women are now promoted proportionally to men), but if that is the case, why the reluctance to provide the data to prove it? When the US Department of Labor conducted an analysis of Google's pay practices in 2017 it found 'systemic compensation disparities against women pretty much across the entire workforce', with 'six to seven standard deviations between pay for men and women in nearly every job category'.[69] Google has since repeatedly refused to hand over fuller pay data to the Labor Department, fighting in court for months to avoid the demand. There was no pay imbalance, they insisted.

For a company built almost entirely on data, Google's reluctance to engage here may seem surprising. It shouldn't be. Software engineer Tracy Chou has been investigating the number of female engineers in the US tech industry since 2013 and has found that '[e]very company has some way of hiding or muddling the data'.[70] They also don't seem interested in measuring whether or not their 'initiatives to make the work environment more female-friendly, or to encourage more women to go into or stay in computing', are actually successful. There's 'no way of judging whether they're successful or worth mimicking, because there are no success metrics attached to any of them', explains Chou. And the result is that 'nobody is having honest conversations about the issue'.

It's not entirely clear why the tech industry is so afraid of sex-disaggregated employment data, but its love affair with the myth of meritocracy might have something to do with it: if all you need to get the 'best people' is to believe in meritocracy, what use is data to you? The irony is, if these so-called meritocratic institutions actually valued science over religion, they could make use of the evidence-based solutions that do already exist. For example, quotas, which, contrary to popular misconception, were recently found by a London School of Economics study to 'weed out incompetent men' rather than promote unqualified women.[71]

They could also collect and analyse data on their hiring procedures to see whether these are as gender neutral as they think. MIT did this, and their analysis of over thirty years of data found that women were disadvantaged by 'usual departmental hiring processes', and that 'exceptional women candidates might very well not be found by conventional departmental search committee methods'.[72] Unless search committees specifically asked department heads for names of outstanding female candidates, they may not put women forward. Many women who were eventually hired when special efforts were made to specifically find female candidates would not have applied for the job without encouragement. In line with the LSE findings, the paper also found that standards were not lowered during periods when special effort was made to hire women: in fact, if anything, the women that were hired 'are somewhat more successful than their male peers'.

The good news is that when organisations do look at the data and attempt to act on it, the results can be dramatic. When a European company advertised for a technical position using a stock photo of a man alongside copy that emphasised 'aggressiveness and competitiveness' only 5% of the applicants were women. When they changed the ad to a stock photo of a woman and focused the text on enthusiasm and innovation, the number of women applying shot up to 40%.[73] Digital design company Made by Many found a similar shift when they changed the wording of their ad for a senior design role to focus more on teamwork and user experience and less on bombastic single-minded egotism.[74] The role was the same, but the framing was different – and the number of female applicants more than doubled.

These are just two anecdotes, but there is plenty of evidence that the wording of an ad can impact on women's likelihood to apply for a job. A study of 4,000 job ads found that women were put off from applying for jobs that used wording associated with masculine stereotypes such as 'aggressive', 'ambitious' or 'persistent'.[75] Significantly, women didn't consciously note the language or realise it was having this impact on them. They rationalised the lack of appeal, putting it down to personal reasons – which goes to show that you don't have to realise you're being discriminated against to in fact be discriminated against.

Several tech start-ups have also taken a leaf out of the New York Philharmonic's book and developed blind recruitment systems.[76] GapJumpers gives job applicants mini assignments designed for a specific post, and the top-performing applicants are sent to hiring managers without any identifying information. The result? Around 60% of those selected end up coming from under-represented backgrounds.

While blind recruitment might work for the initial hiring process, it is less easy to see how it could be incorporated into promotions. But there is a solution here too: accountability and transparency. One tech company made managers truly accountable for their decisions on salary increases by collecting data on all their decisions and, crucially, appointing a committee to monitor this data.[77] Five years after adopting this system, the pay gap had all but disappeared.

CHAPTER 5

The Henry Higgins Effect

When Facebook COO Sheryl Sandberg got pregnant for the first time she was working at Google. 'My pregnancy was not easy,' she wrote in her bestselling book *Lean In*. She had morning sickness for the whole nine months. She didn't just develop a bump, her whole body was swollen. Her feet went up two sizes 'turning into odd-shaped lumps I could see only when they were propped up on a coffee table'.

It was 2014, and Google was already a huge company, with a huge car park – one that Sandberg found increasingly difficult to walk across in her swollen state. After months of struggling she finally went to one of Google's founders, Sergey Brin, and 'announced that we needed pregnancy parking [at the front of the building], preferably sooner rather than later'. Brin agreed immediately, 'noting that he had never thought about it before'. Sandberg herself was 'embarrassed' she hadn't realised 'that pregnant women needed reserved parking until I experienced my own aching feet'.

What Google had suffered from until Sandberg became pregnant was a data gap: neither Google's male founders nor Sandberg had ever been pregnant before. As soon as one of them did get pregnant, that data gap was filled. And all the women who got pregnant at the company after that would benefit from it.

It shouldn't have taken a senior woman getting pregnant for Google to fill this data gap: there had been pregnant women working at the company before. Google could – and should – have been proactive in searching that data out. But the reality is that it usually does take a senior woman for problems like this to be fixed. And so, because business leadership is still so dominated by men, modern workplaces are riddled with these kind of gaps, from doors that are too heavy for the average woman to open with ease, to glass stairs and lobby floors that mean anyone below can see up your skirt, to paving that's exactly the right size to catch your heels. Small, niggling issues that aren't the end of the world, granted, but that nevertheless irritate.

Then there's the standard office temperature. The formula to determine standard office temperature was developed in the 1960s around the metabolic resting rate of the average forty-year-old, 70 kg man.[1] But a recent study found that 'the metabolic rate of young adult females performing light office work is significantly lower' than the standard values for men doing the same type of activity. In fact, the formula may overestimate female metabolic rate by as much as 35%, meaning that current offices are on average five degrees too cold for women. Which leads to the odd sight of female office workers wrapped up in blankets in the New York summer while their male colleagues wander around in summer clothes.[2]

These data gaps are all inequitable, not to mention being bad business sense – an uncomfortable workforce is an unproductive workforce. But workplace data gaps lead to a lot worse than simple discomfort and consequent inefficiency. Sometimes they lead to chronic illness. Sometimes, they mean women die.

Over the past hundred years workplaces have, on the whole, got considerably safer. In the early 1900s around 4,400 people in the UK died at work every year.[3] By 2016, that figure had fallen to 137.[4] In the US, around 23,000 people (out of a workforce of 38 million) died at work in 1913.[5] In 2016, 5,190 people died out of a workforce of 163 million.[6] This significant decrease in fatal accidents has largely been the result of unions pressuring employers and governments to improve safety standards; since the 1974 Health and Safety at Work Act, workplace fatalities in the UK have dropped by 85%. But there is a caveat to this good news story. While serious injuries at work have been decreasing for men, there is evidence that they have been increasing among women.[7]

The rise in serious injuries among female workers is linked to the gender data gap: with occupational research traditionally having been focused on male-dominated industries, our knowledge of how to prevent injuries in women is patchy to say the least. We know all about heavy lifting in construction – what the weight limits should be, how it can be done safely. But when it comes to heavy lifting in care work, well, that's just women's work, and who needs training for that?

Béatrice Boulanger didn't get any training.[8] As a home helper for older people, she 'learned everything on the job'. But her duties included a lot of lifting, often of overweight people. One day, as she was helping a woman out of the bath, her shoulder gave way. 'Everything around the joint was crumbling,' she told occupational health magazine *Hazards*. 'The doctors had to cut off the head of my humerus.' Boulanger eventually needed a full shoulder replacement. And she can no longer do her job.

Boulanger is not a one-off. Women working as carers and cleaners can lift more in a shift than a construction worker or a miner.[9] 'We only got a sink upstairs three years ago,' a cleaner at a cultural centre in France told the *Equal Times*.[10] 'Before that, we had to carry buckets of water upstairs, and down again when the water was dirty. Nobody realised.' And unlike the construction workers and miners, these women often don't go home to rest, but instead go home to a second unpaid shift where there is more lifting, more lugging, more crouching and scrubbing.

In her 2018 retrospective of a lifetime spent researching women's occupational health, Karen Messing, a geneticist and professor of biological sciences at the University of Quebec at Montreal, writes that 'there has still been no biomechanics research on the effects of breast size on lifting techniques associated with back pain'[11] despite the fact that engineer Angela Tate of Memorial University alerted scientists to male bias in biomechanical studies back in the 1990s. Messing also points to women's reports of work-related musculoskeletal pain still being treated with scepticism despite accumulating reports that pain systems function differently among women and men.[12] Meanwhile, we've only just noticed that nearly all pain studies have been done exclusively in male mice.

The gender data gap in occupational health is sometimes attributed to the fact that men are more likely than women to die on the job. But while it is true that the most dramatic accidents are still dominated by male workers this isn't the full story, because an accident at work is by no means the only way your job can kill you. In fact, it's not even the most common way your job can kill you – not by a long shot.

Every year, 8,000 people die from work-related cancers.[13] And although most research in this area has been done on men,[14] it's far from clear that men are the most affected.[15] Over the past fifty years, breast-cancer rates in the industrialised world have risen significantly[16] – but a failure to research female bodies, occupations and environments means that the data for exactly what is behind this rise is lacking.[17] 'We know everything about dust disease in miners,' Rory O'Neill, professor of occupational and environmental policy research at the University of Stirling, tells me. 'You can't say the same for exposures, physical or chemical, in 'women's' work.'

This is partly a historical problem. 'For many long-latency diseases, like cancer,' explains O'Neill, 'it can be decades before the pile of bodies gets big enough to reach a conclusion.' We've been counting the bodies in traditional men's jobs – mining, construction – for several generations. Specifically, we've been counting male bodies: when women did work in those industries, or had similar exposures, 'they were often discounted from studies as "confounding factors".' Meanwhile, in most female-dominated industries, the studies simply weren't done at all. So even if we started the studies now, says O'Neill, it would take a working generation before we had any usable data.

But we aren't starting the studies now. Instead, we continue to rely on data from studies done on men as if they apply to women. Specifically, Caucasian men aged twenty-five to thirty, who weigh 70 kg. This is 'Reference Man' and his superpower is being able to represent humanity as a whole. Of course, he does not.

Men and women have different immune systems and different hormones, which can play a role in how chemicals are absorbed.[18] Women tend to be smaller than men and have thinner skin, both of which can lower the level of toxins they can be safely exposed to. This lower tolerance threshold is compounded by women's higher percentage of body fat, in which some chemicals can accumulate.

The result is that levels of radiation that are safe for Reference Man turn out to be anything but for women.[19] Ditto for a whole range of commonly used chemicals.[20] And yet the male-default one-level-to-rule-them-all approach persists.[21] This is made worse by the way chemicals are tested. To start with, chemicals are still usually tested in isolation, and on the basis of a single exposure. But this is not how women tend to encounter them, either at home (in cleaning products and cosmetics), or in the workplace.

In nail salons, where the workforce is almost exclusively female (and often migrant), workers will be exposed on a daily basis to a huge range of chemicals that are 'routinely found in the polishes, removers, gels, shellacs, disinfectants and adhesives that are staples of their work'.[22] Many of these chemicals have been linked to cancer, miscarriages and lung diseases. Some may alter the body's normal hormonal functions. After a shift of paid work many of these women will then go home and begin a second unpaid shift, where they will be exposed to different chemicals that are ubiquitous in common cleaning products.[23] The effects of these chemicals mixing together are largely unknown,[24] although research does indicate that exposure to a mixture of chemicals can be much more toxic than exposure to chemicals on an individual basis.[25]

Most of the research on chemicals has focused on their absorption through the skin.[26] Leaving aside the problem that absorption through thicker male skin may not be the same as for women, skin is by no means the only way women working in nail salons will be absorbing these chemicals. Many of them are extremely volatile, which means that they evaporate into the air at room temperature and can be inhaled – along with the considerable amounts of dust produced when acrylic nails are filed. The research on how this may impact on workers is virtually non-existent.

But the data, although full of gaps, is mounting. Anne Rochon Ford, a women's health researcher, tells me about how they started to realise there might be a problem in Canada. 'One of the central Toronto community health centres that is very close to Chinatown was seeing a lot of women coming into their clinic who had a particular cluster of conditions that are traditionally associated with chemical exposure,' she explains. It turned out they were all nail-salon workers. Several studies of air quality in nail salons have shown that they rarely exceed occupational exposure limits, but these limits are based on data that doesn't account for the impact of chronic, long-term exposure. And this is particularly an issue when it comes to endocrine disrupting chemicals (EDCs) because, unlike most toxins, they can be harmful even at very low concentrations and they are found in a wide range of plastics, cosmetics and cleaners.[27]

EDCs mimic – and therefore can disrupt – reproductive hormones, 'triggering changes in how cells and organs function, with an impact on a diverse array of metabolic, growth, and reproductive processes in the body'.[28] The data on EDCs and their impact on women is limited.[29] But what we do know is enough to give us pause, and should certainly be enough to trigger a full-scale data-collection programme.

EDCs are known to be linked to breast cancer, and several studies have found that cosmetologists are at a particularly elevated risk of Hodgkin's disease, multiple myeloma and ovarian cancer.[30] When occupational health researchers Jim and Margaret Brophy investigated the chemicals used in automotive plastics workplaces (where plastic parts for motorised vehicles are produced) 'we could not find any substances that they were using that *weren't* suspected' to be either a mammary carcinogen, and/or an endocrine disruptor. 'If you're camping or around a campfire and somebody throws in a plastic bottle or a styrofoam cup people run away,' Brophy points out. 'The smell is enough to tell you it's toxic. Well that's what these women are doing on a daily basis. They're working on moulding machines which heat up these plastic pellets which are full of all kinds of EDCs.'

After ten years working in a job where she is exposed either to mammary carcinogens or an EDC, a woman's risk of developing breast cancer increases by 42%. But the Brophys found that after working for ten years in the auto-plastics industry a woman's likelihood of developing breast cancer trebles. 'And if you were under the age of fifty, so premenopausal breast cancer, it was a fivefold excess.' Even a single year of working in this sector was estimated to increase the odds of developing breast cancer by 9%.[31]

The World Health Organization, the European Union and the Endocrine Society have all issued major reports on the dangers of EDCs, with the Endocrine Society in particular linking their use to the significant increase in breast-cancer rates in industrialised countries.[32] And yet in many countries, regulation of EDCs is spotty at best. Phthalates, some of which have demonstrated endocrine-disrupting properties, are chemicals used to make plastics softer. They are found in 'a wide range of products – from children's toys to shower curtains. They are also used in nail polish, perfumes, and skin moisturizers, and can also be found in the outer coating on medicines and in the tubing used in medical devices'.

In Canada, they 'are explicitly regulated only in soft vinyl articles for children; their use in the Canadian cosmetics industry is largely unregulated'. In the EU, as of 2015 EDCs can't be produced unless authorised for a specific purpose – but they are allowed in products imported from abroad. In the US, there are no federal laws that require companies to list ingredients in their cleaning products (in the US women do 70% of household cleaning and make up 89% of home and hotel cleaners – most of whom are ethnic minorities), and a recent report found that even supposedly 'green' cleaning products contain EDCs.[33] When Always menstrual pads were tested in 2014 they were found to include 'a number of chemicals – including styrene, chloroform and acetone – that have been identified as either carcinogens or reproductive and developmental toxins'.[34]

It's clear that we need more and better data about women's exposure to chemicals. We need data that is separated and analysed by sex, and which includes reproductive status.[35] And physical effects need to be measured for women themselves, rather than being restricted to foetuses and newborns, as is all too often currently the case.[36] We need researchers to understand that because of their unpaid workload women often drop in and out of the workforce and work more than one job at a time (which can lead to, in Rory O'Neill's words, 'a cocktail of exposures'), and that this means that research which tracks only a single, current employment is likely to be sporting a significant gender data gap.[37]

There is no doubt that women are dying as a result of the gender data gap in occupational health research. And there is no doubt that we urgently need to start systematically collecting data on female bodies in the workplace. But there is a second strand to this story because, as the stickiness of the myth of meritocracy shows, closing the gender data gap is only step one. The next, and crucial step, is for governments and organisations to actually use that data to shape policy around it. This isn't happening.

In Canada, even where sex-disaggregated data on chemical exposure exists, the government 'continues to apply a mean all-person daily intake for many substances'.[38] In the UK, where around 2,000 women develop shiftwork-related breast cancer every year, 'breast cancer caused by shiftwork isn't on the state-prescribed disease list'.[39] Neither is asbestos related to ovarian cancer, even though it has the International Agency Research on Cancer's top cancer risk ranking and is the most common gynaecological cancer in UK women. In fact, asbestos-related ovarian cancer cases aren't even tracked and counted by the UK's Health and Safety Executive.

Part of the failure to see the risks in traditionally female-dominated industries is because often these jobs are an extension of what women do in the home (although at a larger and therefore more onerous scale). But the data gap when it comes to women in the workplace doesn't only arise in female-dominated industries. As we've seen, even when women worked in male-dominated industries, they were treated as 'confounding factors', and data on female workers went uncollected.

The result is that even in industries with a good historical health and safety record women are still being failed. In the US, where by 2007 there were nearly 1 million female farm operators, 'virtually all tools and equipment on the US market have been designed either for men or for some "average" user whose size, weight, strength etc. were heavily influenced by the average man'.[40] This has led to tools that are too heavy or long; hand tools that are not appropriately balanced; handles and grips that are not appropriately sized or placed (women's hands are on average 0.8 inches shorter than men's); and mechanised equipment that is too heavy or that is difficult to control (for example pedals on tractors being placed too far from the seat).

Little data exists on injuries to women in construction, but the New York Committee for Occupational Safety & Health (NYCOSH) points to a US study of union carpenters which found that women had higher rates of sprains/strains and nerve conditions of the wrist and forearm than men. Given the lack of data it's hard to be sure exactly why this is, but it's a safe bet to put at least some if not all of the higher injury rates amongst women down to 'standard' construction site equipment being designed around the male body.

Wendy Davis, ex-director of the Women's Design Service in the UK, questions the standard size of a bag of cement. It's a comfortable weight for a man to lift – but it doesn't actually have to be that size, she points out. 'If they were a bit smaller then women could lift them.' Davis also takes issue with the standard brick size. 'I've got photographs of my [adult] daughter holding a brick. She can't get her hand round it. But [her husband] Danny's hand fits perfectly comfortably. Why does a brick have to be that size? It doesn't have to be that size.' She also notes that the typical A1 architect's portfolio fits nicely under most men's arms while most women's arms don't reach round it – and again has photos of her daughter and her husband to prove it. NYCOSH similarly notes that 'standard hand tools like wrenches tend to be too large for women's hands to grip tightly'.[41]

Women in the military are also affected by equipment designed around the male body. In the course of my research I came across the impressively named tactile situation awareness system (TSAS): a vest designed for airforce pilots and fitted with thirty-two sensors that vibrate if the pilot needs to correct her position; pilots can sometimes lose track of where they are in space and cannot tell if they are heading up or down. I say her, because a review of 'Tactile Sensitivity and Human Tactile Interfaces' explained that 'The TSAS allows the pilot to always know his orientation with respect to the ground'.[42] The pronoun choice seems relevant given that the review later casually mentions that '[v]ibration is detected best on hairy, bony skin and is more difficult to detect on soft, fleshy areas of the body.' Women make up 20% of the US airforce and given women have breasts and don't tend to have particularly hairy chests, this sounds like it might be something of a problem for them.[43]

Failing to account for female bodies in the military doesn't just result in equipment that doesn't work for women: it can injure them too. Women in the British Army have been found to be up to seven times more likely than men to suffer from musculoskeletal injuries, even if they have 'the same aerobic fitness and strength'. They are ten times more likely than men to suffer from hip and pelvic stress fractures.[44]

The higher rate of female pelvic stress fractures has been related to what I have christened the 'Henry Higgins effect'. In the 1956 musical *My Fair Lady*, phoneticist Henry Higgins is baffled when, after enduring months of his hectoring put-downs, his protégée-cum-victim Eliza Doolittle finally bites back. 'Why can't a woman be more like a man?' he grumbles. It's a common complaint – and one for which the common solution is to fix the women. This is unsurprising in a world where what is male is seen as universal and what is female is seen as 'atypical'.

And the leadership of the British armed forces have historically been a right bunch of Henry Higginses. Until 2013, when three RAF recruits (one of whom had been medically discharged after suffering four pelvic fractures[45]), challenged the practice in court, women in the British armed forces were forced to match male stride length (the average man's stride is 9–10% longer than the average woman's).[46] Since the Australian Army reduced the required stride length for women from thirty inches to twenty-eight inches, pelvic stress fractures in women have fallen in number. And as an added bonus, not forcing women to march in time with men has not, as yet, led to the apocalypse.

The heavy loads soldiers are required to carry may be aggravating the situation, as women's stride length decreases as loads increase, while men's stride length doesn't show 'significant change'.[47] This may go some way towards explaining US research which found that a woman's risk of injury increases fivefold if she is carrying more than 25% of her body weight.[48] If packs were created for women's bodies, heavy loads might not be such a problem, but they haven't been. Women are more likely to find that rucksacks (which 'have been designed primarily based on the anthropometry of men') are unstable, that pistol belts fit poorly, and that pack straps are uncomfortable.[49] Studies suggest that a 'well-padded hip belt allows a better transfer of the load to the hips' so women can use their stronger leg muscles to carry the load[50] – while men's upper body strength is on average 50% higher than women's, the average gap in lower body strength is about half that. Instead, women compensate for packs built around typically male upper body strength by hyperextending their necks and bringing their shoulders farther forward, leading to injury – and a shorter stride length.

It's not just packs that aren't created to accommodate women's bodies. It wasn't until 2011, thirty-five years after women were first admitted to US military academies, that the first uniforms were designed that accounted for women's hips and breasts.[51] The uniforms also included repositioned knee pads to account for women's generally shorter legs, and, perhaps most exciting of all for a general audience, a redesigned crotch: these uniforms reportedly abandoned the 'universal' zippered fly, instead being designed in such a way that women can pee without pulling down their trousers. But even though the existence of female bodies has finally been recognised by the US military, gaps remain: boots designed to accommodate women's typically narrower feet and higher arches were not included in the uniform changes. According to the *Washington Times*, the US Army buys 'different boot styles for hot and cold weather, mountain and desert warfare and the rain'.[52] Just not for the atypical sex.

The peeing issue is a recurring one for women who have to spend any length of time outdoors. In the UK all coastguards are issued with a set of one-piece overalls which they are meant to put on underneath various other pieces of personal protective equipment (PPE) such as foul-weather clothing, life jackets and climbing harnesses. The double zip at the front of the overalls is great if you are a man, but, explained one woman in a 2017 Trades Union Congress (TUC) report, peeing becomes a 'major operation' for women as all the PPE must be stripped off, followed by the overalls themselves.[53] 'As the type of incidents which we are called to regularly involve long searches which can last for many hours,' she explains, 'you can imagine the discomfort which female coastguards end up having to experience as a result. It has been suggested to management that the current overalls should be replaced with a two-piece garment which would allow the trousers to be pulled down without having to remove the top section, and while management have acknowledged the advantage of this idea nothing has so far been done to implement it.'

A female scientist studying climate change in Alaska was also plagued by overalls designed for the male body.[54] The extreme cold means that overalls are the most sensible thing to wear – but, again, these come with a zip. Where there are indoor toilets, this would be inconvenient and require additional time spent taking off clothes from jacket downwards just for a pee. But when there is no indoor toilet, the problem is much more serious as frostbite becomes a concern. The woman in question bought a rubber funnelled approximation of a penis to deal with the problem – and ended up peeing all over herself. Why can't a woman be more like a man?

In the UK, employers are legally required to provide well-maintained PPE to workers free of charge. But most PPE is based on the sizes and characteristics of male populations from Europe and the US. The TUC found that employers often think that when it comes to female workers all they need to do to comply with this legal requirement is to buy smaller sizes.[55] A 2009 survey by the Women's Engineering Society found 74% of PPE was designed for men.[56] A 2016 Prospect Union survey of women working in sectors ranging from the emergency services, to construction, via the energy industry, found that just 29% wore PPE designed for women,[57] while a 2016 TUC report found that 'less than 10% of women working in the energy sector and just 17% in construction currently wear PPE designed for women'.[58] One rail-industry worker summed it up: 'Size small is a) a rarity, b) men's small only.'

This 'unisex approach' to PPE can lead to 'significant problems', cautions the TUC. Differences in chests, hips and thighs can affect the way the straps fit on safety harnesses. The use of a 'standard' US male face shape for dust, hazard and eye masks means they don't fit most women (as well as a lot of black and minority ethnic men). Safety boots can also be a problem. One female police officer told the TUC about trying to get boots designed for female crime scene investigators. 'The PPE boots supplied are the same as those for males,' she explains, 'and the females find them uncomfortable, too heavy, and causing pressure on the Achilles tendons. Our uniform stores refused to address the matter.'

This isn't just about comfort. Ill-fitting PPE hampers women's work – and can, ironically, sometimes itself be a safety hazard. NYCOSH points out that loose clothing and gloves can get caught in machinery, while overly large boots can cause tripping.[59] Of those surveyed for the 2016 *Prospect* survey, 57% reported that their PPE 'sometimes or significantly hampered their work';[60] over 60% said the same in the Women's Engineering Society survey. One rail-industry worker explained that the 'regular' size thirteen gloves she was issued were 'dangerous for climbing on/off locos' and she had complained to her manager. She doesn't reveal how long it took for management to order her gloves that fit, but another woman who had been issued with the standard size thirteens told *Prospect* that it took her two years to convince her manager to order gloves in her size.

A 2017 TUC report found that the problem with ill-fitting PPE was worst in the emergency services, where only 5% of women said that their PPE never hampered their work, with body armour, stab vests, hi-vis vest and jackets all highlighted as unsuitable.[61] This problem seems to be a global one: in 2018 a female police officer in Spain faced disciplinary action for wearing the women's bulletproof jacket she had bought for herself (at a cost of €500), because the standard-issue men's jacket did not fit her.[62] Pilar Villacorta, women's secretary for the United Association of Civil Guards explained to the *Guardian* that the overly large jackets leave female police officers doubly unprotected: they don't cover them properly and they 'make it hard for female officers to reach their guns, handcuffs and telescopic batons'.[63]

When it comes to front-line workers, poorly fitting PPE can prove fatal. In 1997 a British female police officer was stabbed and killed while using a hydraulic ram to enter a flat. She had removed her body armour because it was too difficult to use the ram while wearing it. Two years later a female police officer revealed that she had to have breast-reduction surgery because of the health effects of wearing her body armour. After this case was reported another 700 officers in the same force came forward to complain about the standard-issue protective vest.[64] But although the complaints have been coming regularly over the past twenty years, little seems to have been done. British female police officers report being bruised by their kit belts; a number have had to have physiotherapy as a result of the way stab vests sit on their female body; many complain there is no space for their breasts. This is not only uncomfortable, it also results in stab vests coming up too short, leaving women unprotected. Which rather negates the whole point of wearing one.

CHAPTER 6

Being Worth Less Than a Shoe

It was in 2008 that the big bisphenol A (BPA) scare got serious. Since the 1950s, this synthetic chemical had been used in the production of clear, durable plastics, and it was to be found in millions of consumer items from baby bottles to food cans to main water pipes.[1] By 2008, 2.7 million tons of BPA was being produced globally every year, and it was so ubiquitous that it had been detected in the urine of 93% of Americans over the age of six.[2] And then a US federal health agency came out and said that this compound that we were all interacting with on a daily basis may cause cancer, chromosomal abnormalities, brain and behavioural abnormalities and metabolic disorders. Crucially, it could cause all these medical problems at levels below the regulatory standard for exposure. Naturally, all hell broke loose.

The story of BPA is in some ways a cautionary tale about what happens when we ignore female medical health data. We have known that BPA can mimic the female hormone oestrogen since the mid-1930s. And since at least the 1970s we have known that synthetic oestrogen can be carcinogenic in women: in 1971 diethylstilbestrol (DES) – another synthetic oestrogen which had been prescribed to millions of pregnant women for thirty years – was banned following reports of rare vaginal cancers in young women exposed to DES while in their mothers' wombs.[3] But BPA carried on being used in hundreds of thousands of tons of consumer plastics: by the late 1980s, production of BPA in the United States 'soared to close to a billion pounds per year as polycarbonates found new markets in compact discs, digital versatile discs (DVDs), water and baby bottles, and laboratory and hospital equipment'.[4]

But the story of BPA is not just about gender: it's also about class. Or at least it's about gendered class. Fearing a major consumer boycott, most baby-bottle manufacturers voluntarily removed BPA from their products, and while the official US line on BPA is that it is not toxic, the EU and Canada are on their way to banning its use altogether. But the legislation that we have exclusively concerns consumers: no regulatory standard has ever been set for workplace exposure.[5] 'It was ironic to me,' says occupational health researcher Jim Brophy, 'that all this talk about the danger for pregnant women and women who had just given birth never extended to the women who were *producing* these bottles. Those women whose exposures *far* exceeded anything that you would have in the general environment. There was no talk about the pregnant worker who is on the machine that's producing this thing.'

This is a mistake, says Brophy. Worker health should be a public health priority if only because 'workers are acting as a canary for society as a whole'. If women's breast-cancer rates in the plastics industry were documented and recognised, 'if we cared enough to look at what's going on in the health of workers that use these substances every day', it would have a 'tremendous effect on these substances being allowed to enter into the mainstream commerce'. It would have a 'tremendous effect on public health'.

But we don't care enough. In Canada, where women's health researcher Anne Rochon Ford is based, five women's health research centres that had been operating since the 1990s, including Ford's own, had their funding cut in 2013. It's a similar story in the UK, where 'public research budgets have been decimated', says Rory O'Neill. And so the 'far better resourced' chemicals industry and its offshoots have successfully resisted regulation for years. They have fought government bans and restrictions. They have claimed that certain chemicals have been removed voluntarily when random testing has shown that they are still present. They have dismissed studies and other evidence of the negative health impacts of their products.[6] Between 1997 and 2005, 115 studies were conducted on BPA in labs all over the world; 90% of those funded by government found BPA had effects at exposures at or below the reference dose. Of the eleven studies funded by industry, none reported any effects.[7]

The result is that workplaces remain unsafe. Brophy tells me that the ventilation he found in most auto-plastics factories was limited to 'fans in the ceiling. So the fumes literally pass the breathing zone and head to the roof and in the summertime when it's really hot in there and the fumes become visible, they will open the doors.' It's the same story in Canadian nail salons, says Rochon Ford. 'It's a Wild West here. Anyone can open a nail salon. It's only recently that you even needed a licence.' But even this is 'pretty lax'. There are no ventilation requirements, there are no training requirements. There is no legislation around wearing gloves and masks. And there is nobody following up on the requirements that do exist – unless someone makes a complaint.

But here we run into another difficulty: who is going to make a complaint? Certainly not the women themselves. Women working in nail salons, in auto-plastics factories, in a vast range of hazardous workplaces, are some of the most vulnerable, powerless workers you can find. They are poor, working class, often immigrants who can't afford to put their immigration status at risk. And this makes them ripe for exploitation.

Auto-plastics factories tend not to be part of the big car companies like Ford. They are usually arms-length suppliers, 'who tend to be non-unionised and tend to be able to get away with more employment-standard violations', Rochon Ford tells me. It doesn't help that Windsor, Ontario, the heart of the auto industry in Canada, has one of the highest unemployment rates in the country. The result is that workers know that if they demand better protections the response will be 'Fine, you're out of here. There's ten women outside the door who want your job.' We've heard factory workers tell us this in the exact same words,' says Rochon Ford.

If this sounds illegal, well, it may be. Over the past hundred years or so, a framework of employee rights has been established. They vary from country to country, but they tend to include a right to paid sick and maternity leave, a right to a set number of hours, and protection from unfair and/or sudden dismissal. But these rights only apply if you are an employee. And, increasingly, many workers are not.

In many nail salons, technicians are technically independent contractors. This makes life much easier for the employers: the inherent risk of running a company based on consumer demand is passed on to workers, who have no guaranteed hours and no job security. Not enough customers today? Don't come in and don't get paid. Minor accident? You're out of here, and forget about redundancy pay.

In 2015 the *New York Times* reported the story of manicurist Qing Lin, forty-seven, who splashed some nail-polish remover on a customer's patent Prada sandals.[8] 'When the woman demanded compensation, the $270 her boss pressed into the woman's hand came out of the manicurist's pay', and Lin was fired. 'I am worth less than a shoe,' she said. Lin's story appeared in a *New York Times* investigation of nail salons which revealed 'all manner of humiliation' suffered by workers, including constant video monitoring by owners, verbal, and even physical abuse.[9] Lawsuits filed in New York courts include allegations of sixty-six-hour weeks at $1.50 an hour and no pay at all on slow days in a salon that charged manicurists for drinking the water.

Following the publication of the *New York Times* investigation a licensing system was introduced in New York. Workers there must be paid at least the minimum wage, and nail salons must display a 'bill of rights' in multiple languages.[10] But workers elsewhere in the US, and elsewhere in the world, are less lucky. In the UK, regulation and licensing of nail bars is largely voluntary[11] – which in practice means largely non-existent. A 2017 report described the predominantly female Vietnamese workforce as 'victims of modern slavery'.[12]

Nail salons are the tip of an extremely poorly regulated iceberg when it comes to employers exploiting loopholes in employment law. Zero-hour contracts, short-term contracts, employment through an agency, these have all been enticingly rebranded the 'gig economy' by Silicon Valley, as if they are of benefit to workers. But the gig economy is in fact often no more than a way for employers to get around basic employee rights. Casual contracts create a vicious cycle: the rights are weaker to begin with, which makes workers reticent to fight for the ones they *do* still have. And so those get bent too. In the UK, which has seen one of the fastest growths in precarious work in the EU,[13] TUC research uncovered a work environment that was rife with employers using casual contracts to illegally undermine workers' rights.[14]

Naturally, the impact of what the International Trade Union Confederation (ITUC) has termed the 'startling growth' of precarious work has barely been gender-analysed.[15] The ITUC reports that its feminised impact is 'poorly reflected in official statistics and government policies', because the 'standard indicators and data used to measure developments on labour markets' are not gender-sensitive, and, as ever, data is often not sex-disaggregated, 'making it sometimes difficult to measure the overall numbers of women'. There are, as a result, 'no global figures related to the number of women in precarious work'.

But the regional and sector-specific studies that do exist suggest 'an overrepresentation of women' in precarious jobs. In the UK, the trade union Unison found that by 2014 women made up almost two-thirds of low-paid workers,[16] and many were 'working multiple jobs on precarious contracts to make up lost hours'.[17] According to a recent Fawcett Society report, one in eight British women is employed on a zero-hours contract.[18] In London that figure is nearly one in three.

And although we often think of precarious work as being relegated to the less 'prestigious' end of the job market it increasingly appears in all sectors and at all levels.[19] According to the UK's University and College Union, tertiary education, usually considered an elite profession, is the second highest user of casual labour.[20] The UCU's data is not sex-disaggregated, but according to the UK's Higher Education Statistics Agency,[21] women are more likely than men to be on shorter, fixed-term contracts, and statistics from Germany and Europe show the same.[22]

More broadly, across the EU most of the increase in women's employment over the past decade has been through part-time and precarious work.[23] In Australia, 30% of women are in casual employment, compared to 22% of men, while in Japan, women make up two-thirds[24] of non-regular workers. A Harvard study on the rise of 'alternative work' in America between 2005 and 2015 found that the percentage of women in such work 'more than doubled', meaning that 'women are now more likely than men to be employed in an alternative work arrangement'.[25]

This is a problem because while precarious work isn't ideal for any worker, it can have a particularly severe impact on women. For a start, it is possible that it is exacerbating the gender pay gap: in the UK there is a 34% hourly pay penalty for workers on zero-hours contracts, a 39% hourly pay penalty for workers on casual contracts, and a 20% pay penalty for agency workers – which are on the increase as public services continue to be outsourced.[26] But no one seems interested in finding out how this might be affecting women. An analysis of pay policy in Europe criticises the outsourcing trend for seeming 'to have been implemented with little or no reference to their gender effects'.[27] And existing data suggests that those gender effects are plentiful.

There is, to begin with, 'limited scope for collective bargaining' in agency jobs. This is a problem for all workers, but can be especially problematic for women because evidence suggests that collective bargaining (as opposed to individual salary negotiation) could be particularly important for women – those pesky modesty norms again. As a result, an increase in jobs like agency work that don't allow for collective bargaining might be detrimental to attempts to close the gender pay gap.

But the negative impact of precarious work on women isn't just about unintended side effects. It's also about the weaker rights that are intrinsic to the gig economy. In the UK a female employee is only entitled to maternity leave if she is actually an employee. If she's a 'worker', that is, someone on a short-term or zero-hours contract, she isn't entitled to any leave at all, meaning she would have to quit her job and reapply after she's given birth. A female worker is also only entitled to statutory maternity pay if she has worked for twenty-six weeks in the last sixty-six and if her average wage is at least £116 per week.

And this is where the problems can kick in. Not being entitled to return to her job meant that Holly, a research associate at a UK university, ended up dropping two pay grades after giving birth.[28] Maria, also a university researcher, had her hours suddenly and mysteriously cut in half six weeks before she was due to give birth; conveniently for her employer, the amount of maternity pay she was owed dropped correspondingly. The same thing happened to Rachel, who works in a pub restaurant: her hours suddenly dropped when she told her employer she was pregnant. She now might not even qualify for statutory maternity pay at all.

After giving birth, Maria ended up on a new university contract for just under three hours a week – the only hours on offer. She can, and does, work extra hours to cover staff absences, but the extra hours are often at short notice. And here we run into the second major problem that disproportionately impacts on female workers: unpredictable, last-minute scheduling.

As we've seen, women still do the vast majority of the world's unpaid care work and, particularly when it comes to childcare, this makes irregular hours extremely difficult. This is partly because, in another case of having the data but failing to use it, British childcare provision has not caught up with the reality of how women are working. We know that 75% of UK families on low to middle incomes now work outside standard hours, but most formal childcare is still only available between 8 a.m. and 6 p.m. It must be booked and paid for well in advance, which is difficult if you don't know when you're going to need it. This problem is particularly acute for single parents (90% of whom in the UK are women[29]) a group that has seen a 27% increase in temporary work.[30] And given Britain has one of the highest childcare costs in Europe, it's also an expensive one.[31]

The scheduling issue is being made worse by gender-insensitive algorithms. A growing number of companies use 'just in time' scheduling software, which use sales patterns and other data to predict how many workers will be needed at any one time. They also respond to real-time sales analyses, telling managers to send workers home when consumer demand is slow. 'It's like magic,' the vice president for business development at Kronos, which supplies the software for a number of US chains, told the *New York Times*.[32]

It probably does feel like magic for the companies that use his software to boost profits by shifting the risks of doing business onto their workers. It probably also feels pretty great for the increasing number of managers who are compensated on the efficiency of their staffing. It feels less great, however, for the workers themselves, particularly those with caring responsibilities. Jannette Navarro, a barista at a Starbucks in San Diego, showed the *New York Times* her upcoming algorithm-produced schedule.[33] It involved working until 11 p.m. on the Friday, reporting again at 4 a.m. on Saturday, and then starting again at 5 a.m. on Sunday. She rarely learned her schedule more than three days in advance, causing havoc for her childcare arrangements – and forcing her to put her associate degree in business on hold. It's another example of how the introduction of Big Data into a world full of gender data gaps can magnify and accelerate already-existing discriminations: whether its designers didn't know or didn't care about the data on women's unpaid caring responsibilities, the software has clearly been designed without reference to them.

A Starbucks spokesperson told the *New York Times* that Navarro's experience 'was an anomaly, and that the company provided at least a week's notice of work hours, as well as stable schedules for employees who want them'. But when journalists spoke to current and former workers 'at 17 Starbucks outlets around the country, only two said they received a week's notice of their hours; some got as little as one day'. And although a few cities have introduced laws regulating the minimum advance notice of a shift an employer can give their workers,[34] there is no nationwide regulation in America – nor is there in many other countries, including in the UK. It is not good enough. The work that (mainly) women do (mainly) unpaid, alongside their paid employment is not an optional extra. This is work that society needs to get done. And getting it done is entirely incompatible with just-in-time scheduling designed entirely without reference to it. Which leaves us with two options: either states provide free, publicly funded alternatives to women's unpaid work, or they put an end to just-in-time scheduling.

A woman doesn't need to be in precarious employment to have her rights violated. Women on irregular or precarious employment contracts have been found to be more at risk of sexual harassment[35] (perhaps because they are less likely to take action against a colleague or employer who is harassing them[36]) but as the #MeToo movement washes over social media, it is becoming increasingly hard to escape the reality that it is a rare industry in which sexual harassment isn't a problem.

As ever, there is a data gap. The TUC warns of a 'paucity of up-to-date, quantitative data on sexual harassment in the workplace', a problem that seems to exist worldwide, with official statistics extremely hard to come by. The UN estimates (estimates are all we have) that up to 50% of women in EU countries have been sexually harassed at work.[37] The figure in China is thought to be as high as 80%.[38] In Australia a study found that 60% of female nurses had been sexually harassed.[39]

The extent of the problem varies from industry to industry. Workplaces that are either male-dominated or have a male-dominated leadership are often the worst for sexual harassment.[40] A 2016 study by the TUC found that 69% of women in manufacturing and 67% of women in hospitality and leisure 'reported experiencing some form of sexual harassment' compared to an average of 52%. A 2011 US study similarly found that the construction industry had the highest rates of sexual harassment, followed by transportation and utilities. One survey of senior level women working in Silicon Valley found that 90% of women had witnessed sexist behaviour; 87% had been on the receiving end of demeaning comments by male colleagues; and 60% had received unwanted sexual advances.[41] Of that 60%, more than half had been propositioned more than once, and 65% had been propositioned by a superior. One in three women surveyed had felt afraid for her personal safety.

Some of the worst experiences of harassment come from women whose work brings them into close contact with the general public. In these instances, harassment all too often seems to spill over into violence.

'He picked her up, threw her across the room, pounded her face and there was blood everywhere.'

'This is when he grabbed me and hit me with the glass. I slumped to the ground and he was still pounding me. [...] I fought him all the way down the hall. He put my head through the wall. There was blood on the walls from my elbows, my face.'

If this doesn't sound like just another day in the office for you, be grateful that you're not a health worker. Research has found that nurses are subjected to 'more acts of violence than police officers or prison guards'.[42] In Ontario in 2014, the number of workplace injuries that required time off work from the healthcare sector 'greatly outnumbered those in other sectors surveyed'. A recent US study similarly found that 'healthcare workers required time off work due to violence four times more often than other types of injury'.[43]

Following the research he conducted with fellow occupational health researcher Margaret Brophy, Jim Brophy concluded that the Canadian health sector was 'one of the most toxic work environments that we had ever seen'. For their 2017 paper on the violence faced by Canadian healthcare workers the Brophys held focus groups where 'people would regularly say, "Every day I go into work and I'm confronted with this."' When the Brophys pulled them up on this claim – surely 'every day' was hyperbole, they meant *often*? 'And they would correct us. "No, we mean every day. It's become part of the job."' One worker recalled the time a patient 'got [a] chair above his head', noting that 'the nursing station has been smashed two or three times'. Other patients used bed pans, dishes, even loose building materials as weapons against nurses.

But despite its prevalence, workplace violence in healthcare is 'an under-reported, ubiquitous, and persistent problem that has been tolerated and largely ignored'. This is partly because the studies simply haven't been done. According to the Brophys' research, prior to 2000, violence against healthcare workers was barely on the agenda: when in February 2017 they searched Medline for 'workplace violence against nurses' they found '155 international articles, 149 of which were published from 2000 to the time of the search'.

But the global data gap when it comes to the sexual harassment and violence women face in the workplace is not just down to a failure to research the issue. It's also down to the vast majority of women not reporting.[44] And this in turn is partly down to organisations not putting in place adequate procedures for dealing with the issue. Women don't report because they fear reprisals and because they fear nothing will be done – both of which are reasonable expectations in many industries.[45] 'We scream,' one nurse told the Brophys. 'The best we can do is scream.'

The inadequacy of procedures to deal with the kind of harassment that female workers face is itself likely also a result of a data gap. Leadership in all sectors is male-dominated and the reality is that men do not face this kind of aggression in the way women do.[46] And so, rather like the Google leadership not thinking to put in pregnancy parking, many organisations don't think to put in procedures to deal adequately with sexual harassment and violence. It's another example of how much a diversity of experience at the top matters for everyone – and how much it matters if we are serious about closing the data gap.[47]

The Brophys warn that gender is also 'typically [...] absent in analyses of health sector violence'. This is unfortunate. According to the International Council of Nurses, 'nurses are the healthcare workers most at risk' – and the vast majority of nurses are women. The absence of gender analysis also means that most of the research doesn't factor in the chronic under-reporting of sexual violence: the Brophys found that only 12% of the workers in their study reported it. 'We don't report sexual violence because it happens so frequently,' explained one woman who had been 'grabbed many times'. But an awareness that the official data is 'believed to grossly underestimate the incidence due to widespread under-reporting' just isn't in the literature, Brophy tells me. This meta data gap goes unremarked.

The violence nurses face at work is not helped by traditional hospital design. The long hallways isolate workers, explains Brophy, scattering them far away from each other. 'Those hallways are terrible,' one worker told Brophy. 'You work way over there – and you can't communicate. I would prefer a full roundabout circle.' This would be an improvement, Brophy points out, because it would enable staff to support each other better. 'If the area was rounded, workers wouldn't be off on one end. If there was two people one would hear something going on.' Most nursing stations don't have protective shatterproof barriers or exits behind the desk, leaving nurses vulnerable to attack. Another worker told Brophy about the time her co-worker was sexually assaulted by a patient. '[Th]e inspector recommended that they put glass up. The hospital fought them on it. They said it stigmatises the patients.'

Both the workers Brophy interviewed and the US's Occupational Health and Safety Administration have highlighted several design features of traditional hospitals ('unsecured access/egress;

insufficient heating or cooling; irritating noise levels; unsecured items') that compound the safety issue – all of which could be addressed without stigmatising anyone. Governments could also reverse policies that result in routine understaffing – an issue that Brophy 'heard in every group in every location', with workers identifying wait times as 'a trigger' for violent behaviour directed towards staff. 'If you don't have the staff to immediately address their issue – if they're kept waiting – they are more likely to escalate in their behaviour,' explained one worker.

Redesigning hospital layouts and increasing staffing levels of course don't come cheap – but there's likely a cost argument that could be made given the amount of time off from injuries and stress workers are taking. Unfortunately, this data is not being 'adequately collected', Brophy tells me. But, he continues, 'I can tell you there's not a doubt in my mind that that is a very high stress work environment and that the demand on people and the limited amount of control they have is the perfect scenario for job burnout.'

And then there's the cost implications of training people who then leave the profession, which came up repeatedly in the focus groups the Brophys conducted. 'We had nurses with twenty-five to thirty years' seniority saying "I'm gonna become a cleaner," or "I'm gonna work in the kitchen because I can't deal with it any more. I can't handle the lack of support and the danger and the risk and coming in every day and facing these things and then being negated and unsupported."'

But even without taking this more long-term view there are plenty of lower-cost options, some of them dazzlingly simple. Consistently charting and flagging patient violence; making reporting procedures less onerous – and having supervisors actually read the reports; ensuring alarms make different noises depending on their purpose: '[I]n one instance, the patient call bell, bathroom assist bell, Code Blue for respiratory or cardiac arrest, and staff emergency alarms all made the same sound in the nurse's station' (fans of British 1970s TV will recognise this problem with alarms as the plot of an actual *Fawlty Towers* episode).

Signs making it clear what behaviour is and isn't acceptable would also be inexpensive. 'I notice at the hospital coffee shop they have a sign that says they won't tolerate any type of verbal abuse,' one woman told the Brophys. 'But there's no signs on our units that say that. [...] There is a poster about if you're widowed and lonely, here's a singles website. But you won't put up a violence sign for us?'

Perhaps most staggeringly simple, participants in the Brophys' research 'suggested that they be permitted to have their last names removed from their name tags – at their employer's expense – as a safety measure'. This would avoid incidents such as when a visitor to the hospital told a female worker, 'Very nice to meet you, [her name]. And you know, you shouldn't have your last name on your badge because I can just look you up and find out who you are and where you live.'

Women have always worked. They have worked unpaid, underpaid, underappreciated, and invisibly, but they have always worked. But the modern workplace does not work for women. From its location, to its hours, to its regulatory standards, it has been designed around the lives of men and it is no longer fit for purpose. The world of work needs a wholesale redesign – of its regulations, of its equipment, of its culture – and this redesign must be led by data on female bodies and female lives. We have to start recognising that the work women do is not an added extra, a bonus that we could do without: women's work, paid and unpaid, is the backbone of our society and our economy. It's about time we started valuing it.

PART III

Design

CHAPTER 7

The Plough Hypothesis

It was the Danish economist Ester Boserup who first came up with the plough hypothesis: that societies that had historically used the plough would be less gender equal than those that hadn't. The theory is based on the relative female-friendliness of shifting agriculture (which is done using handheld tools like hoes or digging sticks) versus plough agriculture (usually driven by a powerful animal like a horse or an ox), the idea being that the former is more accessible to women.[1]

This sex difference in accessibility is partly because of the differences between male and female bodies. Ploughing requires 'significant upper body strength, grip strength, and bursts of power, which are needed to either pull the plough or control the animal that pulls it,' and this privileges male bodies.[2] Upper-body mass is approximately 75%[3] greater in men because women's lean body mass tends to be less concentrated in their upper body,[4] and, as a result, men's upper body strength is on average between 40–60%[5] higher than women's (compared to lower-body strength which is on average only 25% higher in men[6]). Women also have on average a 41%

lower grip strength than men,[7] and this is not a sex difference that changes with age: the typical seventy-year-old man has a stronger handgrip than the average twenty-five-year-old woman.[8] It's also not a sex difference that can be significantly trained away: a study which compared 'highly trained female athletes' to men who were 'untrained or not specifically trained' found that their grip strength 'rarely' surpassed the fiftieth percentile of male subjects.[9] Overall, 90% of the women (this time including untrained women) in the study had a weaker grip than 95% of their male counterparts.

But the disparity in the relative female-friendliness of plough versus shifting agriculture is also a result of gendered social roles. Hoeing can be easily started and stopped, meaning that it can be combined with childcare. The same cannot be said for a heavy tool drawn by a powerful animal. Hoeing is also labour intensive, whereas ploughing is capital intensive,[10] and women are more likely to have access to time rather than money as a resource. As result, argued Boserup, where the plough was used, men dominated agriculture and this resulted in unequal societies in which men had the power and the privilege.

According to a 2011 paper, Boserup's hypothesis holds up to scrutiny.[11] Researchers found that descendants of societies that traditionally practised plough agriculture held more sexist views even if they emigrated to other countries. The paper also found that sexist beliefs correlated with the kind of geo-climactic conditions that would favour plough agriculture over shifting agriculture. This suggested that it was the climate rather than pre-existing sexism that dictated the adoption of the plough – which in turn drove the adoption of sexist views.

The plough theory has its detractors. A 2014 analysis of farming in Ethiopia points out that while farming is strongly identified with men in that country (the farmer is male in 'virtually all Amharic folklore'), and ploughing in particular is exclusively male, the upper-body-strength argument doesn't hold there, because they use a lighter plough (although this of course doesn't deal with the capital investment or childcare issues).[12] This analysis also cites a 1979 paper which disputes the theory on the basis that 'even where the plough never was introduced, among South Cushites in particular, still men are the cultivators'.

Are they though? It's hard to say, because the data on who exactly is doing the farming is, yes, you've guessed it, full of gaps. You'll find no end of reports, articles and briefing papers[13] that include some variation on the claim that 'women are responsible for 60–80% of the agricultural labour supplied on the continent of Africa', but little in the way of evidence. This statistic has been traced back to a 1972 United Nations Economic Commission for Africa, and it's not that it is necessarily wrong, it's just that we can't prove it one way or the other, because we lack the data.

This is partly because, given men and women often farm together, it is difficult to accurately determine how much of the labour of either sex goes into producing an end food product. In a United Nations Food and Agriculture Organization (FAO) paper, economist Cheryl Doss points out that it also depends on how we define and value 'food': by caloric value (where staple crops would come out on top), or by monetary value (where coffee might win)? Given women 'tend to be more heavily involved in the production of staple crops', comparing calorific value 'might indicate a significantly higher share being produced by women.'[14]

'Might' is doing a lot of work there, though, because national surveys often don't report on whether farmers are men or women.[15] Even where data *is* sex-disaggregated, careless survey design can lead to an under-reporting of female labour: if women are asked if they do 'domestic duties' *or* 'work', as if they are mutually exclusive (or as if domestic work is not work), they tend to just select 'domestic duties' because that describes the majority of what they do.[16] This gap is then compounded by the tendency to 'emphasize income-generating activities', the result being that they often underestimate (often female-dominated) subsistence production. The censuses also tend to define agriculture as 'field work', which leads to an under-counting of the women's work 'such as rearing small livestock, kitchen gardening, and post-harvest processing'. It's a fairly clear example of male bias leading to a substantial gender data gap.

A similar problem arises with the division of work by researchers into 'primary' and 'secondary' activities. For a start, secondary activities are not always collected by surveys. Even when they are, they aren't always counted in labour-force figures, and this is a male bias that makes women's paid work invisible.[17] Women will often list their paid work as their secondary activity, simply because their unpaid work takes up so much time, but that doesn't mean that they aren't spending a substantial proportion of their day on paid work. The result is that labour-force statistics often sport a substantial gender data gap.[18]

This male bias is present in the data Doss uses to check the 60–80% statistics. Doss concludes that women make up less than half of the global agricultural labour force, but in the FAO data she uses, 'an individual is reported as being in the agricultural labor force if he or she reports that agriculture is his or her main economic activity'. Which, as we've seen, is to exclude a substantial chunk of women's paid labour. To be fair to Doss, she does acknowledge the issues associated with this approach, critiquing the absurdly low 16% reported share of the agricultural labour force for women in Latin America. Rural women in Latin America, notes Doss, 'are likely to reply that "their home" is their primary responsibility, even if they are heavily engaged in agriculture'.

But even if we were to address all these gender data gaps in calculating female agricultural labour we still wouldn't know exactly how much of the food on your table is produced by women. And this is because female input doesn't equal male output: women on the whole are less productive in agriculture than men. This doesn't mean that they don't work as hard. It means that for the work that they do, they produce less, because agriculture (from tools to scientific research, to development initiatives) has been designed around the needs of men. In fact, writes Doss, given women's various constraints (lack of access to land, credit and new technologies as well as their unpaid work responsibilities) 'it would be surprising if they were able to produce over half of food crops'.

The FAO estimates that if women had the same access to productive resources as men, yields on their farms could increase by up to 30%.[19] But they don't. In an echo of the introduction of the plough, some modern 'labour-saving' devices might more precisely be labelled 'male labour-saving' devices. A 2014 study in Syria, for example, found while the introduction of mechanisation in farming did reduce demand for male labour, freeing men up to 'pursue better-paying opportunities outside of agriculture', it actually increased demand 'for women's labour-intensive tasks such as transplanting, weeding, harvesting and processing'.[20] Conversely, when some agricultural tasks were mechanised in Turkey, women's participation in the agricultural labour force decreased, 'because of men's appropriation of machinery', and because women were reluctant to adopt it. This was in part due to lack of education and sociocultural norms, but also 'because the machinery was not designed for use by women'.[21]

It's not just physical tools that can benefit men at the expense of women. Take what are called 'extension services' (educational programmes designed to teach farmers science-based practice so they can be more productive). Historically, extension services have not been female-friendly. According to a 1988–9 FAO survey (limited to those countries that actually had sex-disaggregated data) only 5% of all extension services were directed towards women.[22] And while things have slightly improved since then,[23] there are still plenty of contemporary examples of development initiatives that forget to include women[24] – and therefore at best don't help, and at worst actively disadvantage them.

A 2015 analysis by Data2x (a UN-backed organisation set up by Hillary Clinton that is lobbying to close the global gender data gap) found that many interventions simply don't reach women in part because women are already overworked and don't have time to spare for educational initiatives, no matter how beneficial they may end up being.[25] Development planners also have to factor in women's (lack of) mobility, in part because of their care responsibilities, but also because they are less likely to have access to transport and often face barriers to travelling alone.

Then there's the language and literacy barrier: many programmes are conducted in the national language, which women are less likely than men to have been taught. Due to the low global levels of female education, women are also less likely to be able to read, so written materials don't help either. These are all fairly basic concerns and shouldn't be hard to account for, but there is plenty of evidence that they continue to be ignored.[26]

Many development initiatives exclude women by requiring a minimum land size, or that the person who attends the training is the head of a farming household, or the owner of the land that is farmed. Others exclude women by focusing solely on farms that have enough money to be able to purchase technology, for example. These conditions are all biased towards male farmers because women dominate the ranks of poor farmers, they dominate the ranks of small-scale farmers, and they are overwhelmingly unlikely to own the land that they farm.[27]

In order to design interventions that actually help women, first we need the data. But it sometimes feels like we're not even trying to collect it. A 2012 Gates Foundation document tells the story of an unnamed organisation that aimed to breed and distribute improved varieties of staple crops.[28] But 'improved' is in the eye of the farmer, and when this organisation did its field-testing it spoke almost exclusively to men. Male farmers said that yield was the most important trait, and so that was the crop that the organisation bred. And then it was surprised when households didn't adopt it.

The decision to talk only to men was bizarre. For all the gaps in our data we can at least say that women do a fair amount of farming: 79% of economically active women in the least developed countries, and 48% of economically active women in the world, report agriculture as their primary economic activity.[29] And the female farmers in this area didn't see yields as the most important thing. They cared about other factors like how much land preparation and weeding these crops required, because these are female jobs. And they cared about how long, ultimately, the crops would take to cook (another female job). The new, high-yield varieties increased the time the women had to spend on these other tasks, and so, unsurprisingly, they did not adopt these crops.

The only thing that development planners need to do to avoid such pitfalls is speak to some women, but they seem bafflingly resistant to this idea. And if you think the decision to design a new staple crop without talking to women is bad, wait until you hear about the history of 'clean' stoves in the developing world.

Humans (by which I mean mainly women) have been cooking with three-stone fires since the Neolithic era. These are exactly what they sound like: three stones on the ground on which to balance a pot, with fuel (wood or whatever else you can gather that will burn) placed in the middle. In South Asia, 75% of families are still using biomass fuels (wood and other organic matter) for energy;[30] in Bangladesh, the figure is as high as 90%.[31] In sub-Saharan Africa biomass fuels are the primary source of energy used for cooking for 753 million people.[32] That's 80% of the population.

The trouble with traditional stoves is that they give off extremely toxic fumes. A woman cooking on a traditional stove in an unventilated room is exposed to the equivalent of more than a hundred cigarettes a day.[33] According to a 2016 paper, in countries from Peru to Nigeria, toxic fumes from stoves are between twenty and a hundred times above World Health Organization guideline limits,[34] and globally they cause three times more deaths (2.9 million)[35] every year than malaria.[36] This is all made worse by the inefficiency of traditional stoves: women who cook on them are exposed to these fumes for three to seven hours a day,[37] meaning that, worldwide, indoor air pollution is the single largest environmental risk factor for female mortality and the leading killer of children under the age of five.[38] Indoor air pollution is also the eighth-leading contributor to the overall global disease burden, causing respiratory and cardiovascular damage, as well as increased susceptibility to infectious illnesses such as tuberculosis and lung cancer.[39] However, as is so often the case with health problems that mainly affect women, 'these adverse health effects have not been studied in an integrated and scientifically rigorous manner'.[40]

Development agencies have been trying to introduce 'clean' stoves since the 1950s, with varying levels of success. The initial impetus was to address deforestation[41] rather than to ease women's unpaid labour or to address the health implications of traditional stove fumes. When it transpired that the environmental disaster was in fact driven by clearing land for agriculture rather than by women's collection of fuel, most of the development industry simply dropped their clean-stove distribution initiatives. Emma Crewe, an anthropologist at SOAS University of London, explains that clean stove initiatives were 'deemed to be a failure as a solution to the energy crisis, and not relevant to any other development area'.[42]

But clean stoves are back on the agenda, and in September 2010 Hillary Clinton announced the formation of the Global Alliance for Clean Cookstoves, which calls for 100 million additional homes to adopt clean and efficient stoves and fuels by 2020.[43] This is a laudable aim, but if it is to be implemented, and if women are actually to use the stoves, a lot of work still remains to be done, not least on data collection.

A 2014 UN publication notes that, relative to data on water and sanitation, country data on access to efficient cookstoves is 'sparse', with national energy policies and poverty reduction strategy papers tending to focus on electrification instead.[44] According to a 2005 World Bank report when it comes to collecting data on people's access to energy governments also tend to measure things like the number of new grid connections, rather than the socio-economic impact of development projects.[45] They also don't generally collect data on what user needs actually are (for example, drinking-water pumping; food processing; fuel collection) before starting on their development projects. And the result of this dearth of data is that, to date, clean cook stoves have nearly all been rejected by users.

In the 1990s Emma Crewe was informed by stove technicians that low adoption was because users came from a 'conservative culture'.[46] They needed 'educating' in proper stove usage. Women are still being blamed in the twenty-first century. A 2013 WASHplus- and USAID-funded report on user experiences of five stoves in Bangladesh repeatedly acknowledged that all five stoves increased cooking time and required more attending.[47] This prevented women from multitasking as they would with a traditional stove, and forced them to change the way they cooked – again increasing their workload. Nevertheless, the main and repeated recommendation of the report was to fix the women, rather than the stoves. The women needed to be educated on how great the 'improved' stoves were, rather than stove designers needing to be educated on how not to increase women's already fifteen-hour average working day.[48]

Despite what academics, NGOs and expatriate technicians seem to think, the problem is not the women. It is the stoves: developers have consistently prioritised technical parameters such as fuel efficiency over the needs of the stove user, frequently leading users to reject them, explains Crewe.[49] And although the low adoption rate is a problem going back decades, development agencies have yet to crack the problem,[50] for the very simple reason that they still haven't got the hang of consulting women and *then* designing a product rather than enforcing a centralised design on them from above.[51]

One Indian programme failed because while the new stove worked well in the lab, it required more maintenance than traditional stoves – maintenance the designers had simply assumed the 'household' would take care of.[52] But structural repairs in Orissa are traditionally the responsibility of men, who didn't see fixing the new stoves as a priority, because their wives could still prepare meals using the traditional stoves. So the women went back to using the toxic fume-producing traditional stoves, while the new stoves gathered dust in corners.

The issue of gendered priorities also affects household spending and therefore, if a household will adopt a stove at all. Despite hundreds of attempts to introduce a variety of clean stoves in Bangladesh since the early 1980s, over 98% of the rural population continue to cook with traditional biomass-burning stoves.[53] A 2010 study which set out to understand why, found that women 'seemed to exhibit a stronger preference than men for any improved stove, in particular for the health-saving chimney stoves', and were more likely to order stoves when asked without their husbands present. But when the team returned to deliver the stoves four months later, the gender gap had disappeared; women's preferences had fallen back into line with their husbands'.

That women's failure to adopt clean stoves may simply result from a lack of purchasing authority is backed up by a 2016 report which found that 'female-headed households are more likely to adopt cleaner cooking solutions than male-headed households'.[54] Meanwhile a 2012 Yale study found that 94% of respondents 'believed that indoor smoke from the traditional stoves is harmful', but 'opted for traditional cookstove technology so they could afford basic needs' – although this didn't prevent the university from headlining a press release on the study 'Despite efforts for change, Bangladeshi women prefer to use pollution-causing cookstoves,' as if the women were perverse rather than lacking in purchasing authority.[55] Perhaps silly women obstreperously choosing air pollution for no good reason made for a better headline than endemic poverty.

This decades-long failure to design either stoves or implementation plans that account for women's needs is a health disaster that is set to get worse. As climate change makes high-quality fuel increasingly scarce (because of soil erosion and desertification), women are forced to use leaves, straw and dung, which give off fumes that are even more toxic. And this is a travesty because there is no doubt that clean stoves would significantly improve women's lives. A 2011 Yemen study found that women who lacked access to water and gas stoves spent 24% of their time engaged in paid work; this rose to around 52% for women who did have access.[56] A 2016 report into stove use in India found that when women did adopt clean stoves (for example the cheap and portable Anagi 2 which has been found to substantially decrease cooking time), they had more time for social and family activities and community meetings.[57] Households with clean stoves also reported sending their children to school more often.[58]

There is some cause for hope. In November 2015, researchers in India reported[59] that they had conducted a successful field study using 'an inexpensive (USD $1) device that may be simply placed in existing three-stone hearths'. This simple device cut wood use and smoke 'to levels comparable to those achieved by the more expensive high-efficiency cookstoves'. This breakthrough came about as a result of filling a decades-long data gap: noting that the two decades of government attempts to implement high-efficiency cookstoves (HECs) in rural India had been largely unsuccessful, the researchers decided to investigate why.

And by speaking to women, they found out: HECs were unable to accept 'large pieces of wood without having them split lengthwise', an issue also uncovered in the 2013 study of five clean stoves mentioned earlier. These researchers understood that everything to do with cooking, including fuel, was the domain of women, and that since splitting wood was 'very difficult for the women to do', it was perfectly rational for women to 'abandon these HECs since their traditional *chulhas* (mud and brick stoves) have no such size limitation'.

Based on their findings they set about fixing the stove technology to fit the women. Realising that 'a single HEC stove cannot possibly replace all of these traditional stoves', the researchers concluded that 'significant fuelwood reductions can only be achieved with locally customizable solutions in different parts of the world'. The result of their data-led design was the *mewar angithi* (MA), a simple metal device that 'was engineered to be placed in a traditional *chulha* in order to provide the same airflow mechanism in the traditional *chulha* as occurs in the HEC stoves'.

To keep costs down (another regular concern of stove users), they constructed the device from metal washer industry scrap metal that they found in a local market 'at one-fourth the cost of solid metal sheets'. And because of the 'simple, bent plate design of the MA, it is easily customized to individual *chulha* units'. Since then, studies in Kenya[60] and Ghana[61] with the same device have found similarly positive results, showing what can be achieved when designers start from the basis of closing the gender data gap.

CHAPTER 8

One-Size-Fits-Men

In 1998, a pianist called Christopher Donison wrote that 'one can divide the world into roughly two constituencies': those with larger hands, and those with smaller hands. Donison was writing as a male pianist who, due to his smaller than average hands, had struggled for years with traditional keyboards, but he could equally have been writing as a woman. There is plenty of data showing that women have, on average, smaller hands than men,[1] and yet we continue to design equipment around the average male hand as if one-size-fits-men is the same as one-size-fits-all.

This one-size-fits-men approach to supposedly gender-neutral products is disadvantaging women. The average female handspan is between seven and eight inches,[2] which makes the standard forty-eight-inch keyboard something of a challenge. Octaves on a standard keyboard are 7.4 inches wide, and one study found that this keyboard disadvantages 87% of adult female pianists.[3] Meanwhile, a 2015 study which compared the handspan of 473 adult pianists to their 'level of acclaim' found that all twelve of the pianists considered to be of international renown had spans of 8.8 inches or above.[4]

Of the two women who made it into this exalted group, one had a handspan of nine inches and the other had a handspan of 9.5 inches.

The standard piano keyboard doesn't just make it harder for female pianists to match the level of acclaim reached by their male colleagues: it also affects their health. A range of studies carried out on instrumentalists during the 1980s and 90s found that female musicians suffered 'disproportionately' from work-related injuries, and that keyboard players were among those 'most at risk'. Several studies have found that female pianists run an approximately 50% higher risk of pain and injury than male pianists; in one study 78% of women compared to 47% of men had developed RSI.[5]

It seems likely that this is related to hand size: another study from 1984, which included only male pianists, identified twenty-six 'successful performers' defined as 'well-known soloists and winners of international competitions', and ten 'problem cases': those who had 'struggled with technical or injury problems over a long period'.[6] The former group's average handspan was 9.2 inches compared to the problem cases' 8.7 inches – which is nevertheless still substantially larger than the average female handspan.

It was while Christopher Donison was practising the coda of the G minor Chopin Ballade on his Steinway concert grand 'for about the thousandth time', that he had the thought that led to his designing a new keyboard for people with smaller hands. What if it wasn't that his hands were too small, but that the standard keyboard was too large? The result of this thought was the 7/8 DS keyboard, which, Donison claimed, transformed his playing. 'I could finally use the correct fingerings. Broken-chord formations could be played on one hand position, instead of two. [...] Wide, sweeping, left-hand arpeggiated figures so prevalent in Romantic music become possible, and I could actually get on with the business of cultivating the right sound, rather than repeatedly practicing the same passage.'[7] Donison's experience is backed up by numerous studies which have also found that a 7/8 keyboard dispels the professional and health disadvantages imposed by the conventional keyboard.[8] And yet there remains a strange (that is, if you don't accept that sexism is at play here) reluctance in the piano world to adapt.

The reluctance to abandon design that suits only the largest male hands seems endemic. I remember a time back in the early 2000s when it was the smallest handsets that were winning phone-measuring contests. That all changed with the advent of the iPhone and its pretenders. Suddenly it was all about the size of your screen, and bigger was definitely better. The average smartphone is now 5.5 inches,[9] and while we're admittedly all extremely impressed by the size of your screen, it's a slightly different matter when it comes to fitting into half the population's hands (not to mention minuscule or non-existent pockets). The average man can fairly comfortably use his device one-handed – but the average woman's hand is not much bigger than the handset itself.

This is obviously annoying – and foolish for a company like Apple, given that research shows women are more likely to own an iPhone than men.[10] But don't expect to uncover a method to their madness any time soon, because it's extraordinarily difficult to get any smartphone company to comment on their massive-screen fixation. In desperation for answers I turned to the *Guardian*'s tech reporter Alex Hern. But he couldn't help me either. 'It's a noted issue,' he confirmed, but 'one I've never got a straight answer on.' Speaking to people informally, he said, the 'standard response' was that phones were no longer designed for one-handed use. He's also been told that actually many women opt for larger phones, a trend that was 'usually attributed to handbags'. And look, handbags are all well and good, but one of the reasons women carry them in the first place is because our clothes lack adequate pockets. So designing phones to be handbag-friendly rather than pocket-friendly feels like adding injury (more on this later) to insult. In any case, it's rather odd to claim that phones are designed for women to carry in their handbags when so many passive-tracking apps clearly assume your phone will be either in your hands or in your pockets at all times, rather than sitting in your handbag on your office desk.

I next turned to award-winning tech journalist and author James Ball, who has another theory for why the big-screen fixation persists: because the received wisdom is that men drive high-end smartphone purchases, women in fact don't figure in the equation at all. If this is true it's certainly an odd approach for Apple to take given the research about women being more likely to own iPhones. But I have another, more fundamental complaint with this analysis, because it again suggests that the problem is with women, rather than male-biased design. In other words: if women aren't driving high-end smartphone purchases is it because women aren't interested in smartphones, or could it be because smartphones are designed without women in mind? On the bright side, however, Ball reassured me that screens probably wouldn't be getting any bigger because 'they've hit the limit of men's hand size'.

Good news for men, then. But tough breaks for women like my friend Liz who owns a third-generation Motorola Moto G. In response to one of my regular rants about handset sizes she replied that she'd just been 'complaining to a friend about how difficult it was to zoom on my phone camera. He said it was easy on his. Turns out we have the same phone. I wondered if it was a hand-size thing'.

Almost certainly, it was. When Zeynep Tufekci, a researcher at the University of North Carolina, was trying to document tear-gas use in the Gezi Park protests in Turkey in 2013, the size of her Google Nexus got in the way.[11] It was the evening of 9 June. Gezi Park was crowded. Parents were there with their children. And then the canisters were fired. Because officials 'often claimed that tear gas was used only on vandals and violent protesters', Tufekci wanted to document what was happening. So she pulled out her phone. 'And as my lungs, eyes and nose burned with the pain of the lachrymatory agent released from multiple capsules that had fallen around me, I started cursing.' Her phone was too big. She could not take a picture one-handed – 'something I had seen countless men with larger hands do all the time'. All Tufekci's photos from the event were unusable, she wrote, and 'for one simple reason: good smartphones are designed for male hands'.

Like the standard keyboard, smartphones designed for male hands also may be affecting women's health. It is a relatively new field of study, but the research that does exist on the health impact of smartphones is not positive.[12] But although women's hand size is demonstrably smaller than men's, and although women have been found to have a higher prevalence of musculoskeletal symptoms and disorders,[13] research into the impact of large smartphones on hands and arms does not buck the gender data gap trend. In the studies I found, women were significantly under-represented as subjects,[14] and the vast majority of studies did not sex-disaggregate their data[15] – including those that did manage to adequately represent women.[16] This is unfortunate because the few that did sex-disaggregate their data reported a statistically significant gender difference in the impact of phone size on women's hand and arm health.[17]

The answer to the problem of smartphones that are too big for women's hands seems obvious: design smaller handsets. And there are of course some smaller handsets on the market, notably Apple's iPhone SE. But the SE wasn't updated for two years and so was an inferior product to the standard iPhone range (which offers only huge or huger as size options). And it's now been discontinued anyway. In China, women and men with smaller hands can buy the Keecoo K1 which, with its hexagonal design, is trying to account for women's hand size: good.[18] But it has less processing power and comes with in-built air-brushing: bad. Very bad.

Voice recognition has also been suggested as a solution to smartphone-associated RSI,[19] but this actually isn't much of a solution for women, because voice-recognition software is often hopelessly male-biased. In 2016, Rachael Tatman, a research fellow in linguistics at the University of Washington, found that Google's speech-recognition software was 70% more likely to accurately recognise male speech than female speech[20] – and it's currently the best on the market.[21]

Clearly, it is unfair for women to pay the same price as men for products that deliver an inferior service to them. But there can also be serious safety implications. Voice-recognition software in cars, for example, is meant to decrease distractions and make driving safer. But they can have the opposite effect if they don't work – and often, they don't work, at least for women. An article on car website *Autoblog* quoted a woman who had bought a 2012 Ford Focus, only to find that its voice-command system only listened to her husband, even though he was in the passenger seat.[22] Another woman called the manufacturer for help when her Buick's voice-activated phone system wouldn't listen to her: 'The guy told me point-blank it wasn't ever going to work for me. They told me to get a man to set it up.' Immediately after writing these pages I was with my mother in her Volvo Cross-Country watching her try and fail to get the voice-recognition system to call her sister. After five failed attempts I suggested she tried lowering the pitch of her voice. It worked first time.

As voice-recognition software has become more sophisticated, its use has branched out to numerous fields, including medicine, where errors can be just as grave. A 2016 paper analysed a random sample of a hundred notes dictated by attending emergency physicians using speech-recognition software, and found that 15% of the errors were critical, 'potentially leading to miscommunication that could affect patient care'.[23] Unfortunately these authors did not sex-disaggregate their data, but papers that have, report significantly higher transcription error rates for women than men.[24] Dr Syed Ali, the lead author of one of the medical dictation studies, observed that his study's 'immediate impact' was that women 'may have to work somewhat harder' than men 'to make the [voice recognition] system successful'.[25] Rachael Tatman agrees: 'The fact that men enjoy better performance than women with these technologies means that it's harder for women to do their jobs. Even if it only takes a second to correct an error, those seconds add up over the days and weeks to a major time sink, time your male colleagues aren't wasting messing with technology.'

Thankfully for frustrated women around the world, Tom Schalk, the vice president of voice technology at car navigation system supplier ATX, has come up with a novel solution to fix the 'many issues with women's voices'.[26] What women need, he said, was 'lengthy training' – if only women 'were willing' to submit to it. Which, sighs Schalk, they just aren't. Just like the wilful women buying the wrong stoves in Bangladesh, women buying cars are unreasonably expecting voice-recognition software developers to design a product that works for them when it's obvious that the problem needing fixing is the women themselves. Why can't a woman be more like a man?

Rachael Tatman rubbishes the suggestion that the problem lies in women's voices rather than the technology that doesn't recognise them: studies have found that women have 'significantly higher speech intelligibility',[27] perhaps because women tend to produce longer vowel sounds[28] and tend to speak slightly more slowly than men.[29] Meanwhile, men have 'higher rates of disfluency, produce words with slightly shorter durations, and use more alternate ('sloppy') pronunciations'.[30] With all this in mind, voice-recognition technology should, if anything, find it easier to recognise female rather than male voices – and indeed, Tatman writes that she has 'trained classifiers on speech data from women and they worked just fine, thank you very much'.

Of course, the problem isn't women's voices. It's our old friend, the gender data gap. Speech-recognition technology is trained on large databases of voice recordings, called corpora. And these corpora are dominated by recordings of male voices. As far as we can tell, anyway: most don't provide a sex breakdown on the voices contained in their corpus, which in itself is a data gap of course.[31] When Tatman looked into the sex ratio of speech corpora only TIMIT ('the single most popular speech corpus in the Linguistic Data Consortium') provided data broken down by sex. It was 69% male. But contrary to what these findings imply, it is in fact possible to find recordings of women speaking: according to the data on its website, the British National Corpus (BNC)[32] is sex-balanced.[33]

Voice corpora are not the only male-biased databases we're using to produce what turn out to be male-biased algorithms. Text corpora (made up of a wide variety of texts from novels, to newspaper articles, to legal textbooks) are used to train translation software, CV-scanning software, and web search algorithms. And they are riddled with gendered data gaps. Searching the BNC[34] (100 million words from a wide range of late twentieth-century texts) I found that female pronouns consistently appeared at around half the rate of male pronouns.[35] The 520-million-word Corpus of Contemporary American English (COCA) also has a 2:1 male to female pronoun ratio despite including texts as recent as 2015.[36] Algorithms trained on these gap-ridden corpora are being left with the impression that the world actually *is* dominated by men.

Image datasets also seem to have a gender data gap problem: a 2017 analysis of two commonly used datasets containing 'more than 100,000 images of complex scenes drawn from the web, labeled with descriptions' found that images of men greatly outnumber images of women.[37] A University of Washington study similarly found that women were under-represented on Google Images across the forty-five professions they tested, with CEO being the most divergent result: 27% of CEOs in the US are female, but women made up only 11% of the Google Image search results.[38] Searching for 'author' also delivered an imbalanced result, with only 25% of the Google Image results for the term being female compared to 56% of actual US authors, and the study also found that, at least in the short term, this discrepancy did affect people's views of a field's gender proportions. For algorithms, of course, the impact will be more long term.

As well as under-representing women, these datasets are misrepresenting them. A 2017 analysis of common text corpora found that female names and words ('woman', 'girl', etc.) were more associated with family than career; it was the opposite for men.[39] A 2016 analysis of a popular publicly available dataset based on Google News found that the top occupation linked to women was 'homemaker' and the top occupation linked to men was 'maestro'.[40] Also included in the top ten gender-linked occupations were philosopher, socialite, captain, receptionist, architect and nanny – I'll leave it to you to guess which were male and which were female. The 2017 image dataset analysis also found that the activities and objects included in the images showed a 'significant' gender bias.[41] One of the researchers, Mark Yatskar, saw a future where a robot trained on these datasets who is unsure of what someone is doing in the kitchen 'offers a man a beer and a woman help washing dishes'.[42]

These cultural stereotypes can be found in artificial intelligence (AI) technologies already in widespread use. For example, when Londa Schiebinger, a professor at Stanford University, used translation software to translate a newspaper interview with her from Spanish into English, both Google Translate and Systran repeatedly used male pronouns to refer to her, despite the presence of clearly gendered terms like '*profesora*' (female professor).[43] Google Translate will also convert Turkish sentences with gender-neutral pronouns into English stereotypes. '*O bir doktor*,' which means 'S/he is a doctor' is translated into English as 'He is a doctor', while '*O bir hemşire*' (which means 'S/he is a nurse') is rendered 'She is a nurse'. Researchers have found the same behaviour for translations into English from Finnish, Estonian, Hungarian and Persian.

The good news is that we now have this data – but whether or not coders will use it to fix their male-biased algorithms remains to be seen. We have to hope that they will, because machines aren't just reflecting our biases. Sometimes they are amplifying them – and by a significant amount. In the 2017 images study, pictures of cooking were over 33% more likely to involve women than men, but algorithms trained on this dataset connected pictures of kitchens with women 68% of the time. The paper also found that the higher the original bias, the stronger the amplification effect, which perhaps explains how the algorithm came to label a photo of a portly balding man standing in front of a stove as female. Kitchen > male pattern baldness.

James Zou, assistant professor of biomedical science at Stanford, explains why this matters. He gives the example of someone searching for 'computer programmer' on a program trained on a dataset that associates that term more closely with a man than a woman.[44] The algorithm could deem a male programmer's website more relevant than a female programmer's – 'even if the two websites are identical except for the names and gender pronouns'. So a male-biased algorithm trained on corpora marked by a gender data gap could literally do a woman out of a job.

But web search is only scraping the surface of how algorithms are already guiding decision-making. According to the *Guardian* 72% of US CVs never reach human eyes,[45] and robots are already involved in the interview process with their algorithms trained on the posture, facial expressions and vocal tone of 'top-performing employees'.[46] Which sounds great – until you start thinking about the potential data gaps: did the coders ensure that these top-performing employees were gender and ethnically diverse and, if not, does the algorithm account for this? Has the algorithm been trained to account for socialised gender differences in tone and facial expression? We simply don't know, because the companies developing these products don't share their algorithms – but let's face it, based on the available evidence, it seems unlikely.

AI systems have been introduced to the medical world as well, to guide diagnoses – and while this could ultimately be a boon to healthcare, it currently feels like hubris.[47] The introduction of AI to diagnostics seems to be accompanied by little to no acknowledgement of the well-documented and chronic gaps in medical data when it comes to women.[48] And this could be a disaster. It could, in fact, be fatal – particularly given what we know about machine learning amplifying already-existing biases. With our body of medical knowledge being so heavily skewed towards the male body, AIs could make diagnosis for women worse, rather than better.

And, at the moment, barely anyone is even aware that we have a major problem brewing here. The authors of the 2016 Google News study pointed out that not a single one of the 'hundreds of papers' about the applications for word-association software recognised how 'blatantly sexist' the datasets are. The authors of the image-labelling paper similarly noted that they were 'the first to demonstrate structured prediction models amplify bias and the first to propose methods for reducing this effect'.

Our current approach to product design is disadvantaging women. It's affecting our ability to do our jobs effectively – and sometimes to even get jobs in the first place. It's affecting our health, and it's affecting our safety. And perhaps worst of all, the evidence suggests that when it comes to algorithm-driven products, it's making our world even more unequal. There are solutions to these problems if we choose to acknowledge them, however. The authors of the women = homemaker paper devised a new algorithm that reduced gender stereotyping (e.g. 'he is to doctor as she is to nurse') by over two-thirds, while leaving gender-appropriate word associations (e.g. 'he is to prostate cancer as she is to ovarian cancer') intact.[49] And the authors of the 2017 study on image interpretation devised a new algorithm that decreased bias amplification by 47.5%.

CHAPTER 9

A Sea of Dudes

When Janica Alvarez was trying to raise funds for her tech start-up Naya Health Inc. in 2013, she struggled to get investors to take her seriously. In one meeting, 'investors Googled the product and ended up on a porn site. They lingered on the page and started cracking jokes', leaving Alvarez feeling like she was 'in the middle of a fraternity'.[1] Other investors were 'too grossed out to touch her product or pleaded ignorance', with one male investor saying 'I'm not touching that; that's disgusting.'[2] And what was this vile, 'disgusting' and incomprehensible product Alvarez was pitching? Reader, it was a breast pump.

The odd thing is, the breast-pump industry is one that is ripe for 'disruption', as Silicon Valley would have it. Breast-pumping is huge business in the US in particular: given the lack of legally mandated maternity leave, for most American women breast-pumping is the only option if they want to follow their doctors' recommendations and breastfeed their babies for at least six months (in fact, the American Academy of Pediatrics recommends that women try to breastfeed for at least twelve months).[3]

And one company, Medela, has pretty much cornered the market. According to the *New Yorker*, 'Eighty per cent of hospitals in the United States and the United Kingdom stock Medela's pumps, and its sales increased thirty-four per cent in the two years after the passage of the Affordable Care Act, which mandated coverage of lactation services, including pumps.' But the Medela pump is just not very good. Writing for the *New Yorker*[4] Jessica Winter described it as a 'hard, ill-fitting breast shield with a bottle dangling from it', which, as it sucks milk out of a woman's breast 'pulls and stretches the breast like it's taffy, except that taffy doesn't have nerve endings'.[5] And although some women manage to make it work hands-free most can't because it doesn't fit well enough. So they just have to sit and hold their personal milking contraptions to their breasts, for twenty minutes a time, several times a day.

So, to sum up: captive market (currently estimated at $700 million with room to grow)?[6] Check. Products that aren't serving consumer needs? Check. Why aren't investors lapping it up?

Addressing the under-representation of women in positions of power and influence is often framed as a good in itself. And, of course, it is. It is a matter of justice that women have an equal chance of success as their equally qualified male colleagues. But female representation is about more than a specific woman who does or doesn't get a job, because female representation is also about the gender data gap. As we saw with Sheryl Sandberg's story about pregnancy parking, there will be certain female needs men won't think to cater for because they relate to experiences that men simply won't have. And it's not always easy to convince someone a need exists if they don't have that need themselves.

Dr Tania Boler, founder of women's health tech company Chiaro, thinks that the reluctance to back female-led companies is partly a result of the 'stereotype that men like great design and great tech and women don't'. But is this stereotype based in reality, or is it possible that the problem isn't tech-blind women so much as woman-blind tech, created by a woman-blind tech industry and funded by woman-blind investors?

A substantial chunk of tech start-ups are backed by venture capitalists (VCs) because they can take risks where banks can't.[7] The problem is that 93% of VCs are men,[8] and, 'men back men', explains Debbie Woskow, co-founder of AllBright, a members' club, academy, and fund that backs female-led business. 'We need to have more women writing cheques. And men need to recognise that backing women is a great investment.' Woskow tells me that when she was in the process of setting up AllBright with her friend Anna Jones, the former CEO of Hearst, 'men who should know better, to be honest' would 'frequently' tell them, 'That's lovely, it's great that you and Anna have set up a charity.' Woskow bristles at this. 'We're not a charity. We're doing this because women deliver great economic returns.'

The data suggests she's not wrong. Research published in 2018 by Boston Consulting Group found that although on average female business owners receive less than half the level of investment their male counterparts get, they produce more than twice the revenue.[9] For every dollar of funding, female-owned start-ups generate seventy-eight cents, compared to male-owned start-ups which generate thirty-one cents. They also perform better over time, 'generating 10% more in cumulative revenue over a five-year period'.

This may be partly because women are 'better suited for leadership than men'.[10] That was the conclusion of a study conducted by BI Norwegian Business School, which identified the five key traits (emotional stability, extraversion, openness to new experiences, agreeableness and conscientiousness) of a successful leader. Women scored higher than men in four out of the five. But it may also be because the women who do manage to make it through are filling a gender data gap: studies have repeatedly found that the more diverse a company's leadership is, the more innovative they are.[11] This could be because women are just innately more innovative – but more likely is that the presence of diverse perspectives makes businesses better informed about their customers. Certainly, innovation is strongly linked to financial performance.

And when it comes to consumer electronics for women, Boler says, innovation has been sorely lacking. 'There's never been much innovation in consumer electronics for women,' she says. 'It's always focused on a very superficial aesthetic level: turn something pink, or turn something into a piece of jewellery, rather than taking account of the fact that technology can solve real problems for women.' The result has been a chronic lack of investment, meaning that 'the actual technology that's used in medical devices for women is sort of a kickback from the 1980s'.

When I interview her early in 2018 Tania Boler is about to launch her own breast pump, and she is scathing about what is currently available on the market. 'It's just horrible,' she says, bluntly. 'It's painful, it's loud, it's difficult to use. It's quite humiliating.' I think of trying to hold a conversation with my sister-in-law as she sits on the sofa with her top off, her breasts wired up to a machine. 'It's not even that complicated to get it right,' Boler adds. The notion that 'it would be nice to pump while you're able to do something else, rather than having to spend hours a day sitting there chained to this noisy machine' should, she says be 'a basic requirement'. But somehow, it hasn't been. When I ask her why she thinks this is, Boler muses that perhaps it's different for her because she's a woman. So 'I just go in with: "As a woman what do I want from this?"'

But if the data gap of what women actually want is fairly easily fixed by, well, asking women, there's another more chronic gap: data on the female body itself. Boler developed her first product – Elvie, a smart pelvic-floor trainer – after realising that poor pelvic-floor health in women was 'a massively hidden epidemic':

37% of women suffer from pelvic-floor issues; 10% of women will need to have an operation at some point because of prolapse (where your organs start dropping through your vagina). This rises to 50% of women over fifty.

'There's a sense of injustice,' says Boler. 'It's a big issue for women and it should be a normal part of how women look after their bodies. But you need to have information and data in order to do that.' And when Boler was first researching the issue, that data simply didn't exist. 'We were trying to design a product which fits in the vagina, and so we needed to answer simple questions like, what size, how does it vary by age, by race, after children – all the usual questions. And there just was no data there *at all*. [...] Fifty per cent of the population have a vagina,' she continues, 'and yet there's hardly any journal articles about this part of anatomy. Three years ago I found about four articles done decades ago.' One of them was 'literally by a guy who basically made a kind of plaster cast, like a mould inside the vagina, and concluded that there were four shapes: a mushroom and a cone and a heart ...' she trails off laughing.

Problems with pelvic-floor health are often preventable, and the evidence base for pelvic-floor training is 'very strong', Boler tells me. 'It's the number-one line of defence and it's recommended under the NICE [National Institute for Health and Case Excellence] guidelines in the UK.' But when she started looking at the technology in hospitals, 'there had been no investment. It was so outdated, it was very unreliable and not even very valid.' The current treatment for prolapse (to insert a mesh into the vagina) is the subject of an ongoing scandal in the UK, as hundreds of women have been left in severe, debilitating pain, by what they describe as 'barbaric' treatment.[12] In Scotland, a woman just died.

Ida Tin, founder of menstrual-tracking app Clue, encountered the same problem when she started trying to find an alternative to traditional contraception. 'Menstruation is listed as one of the vital signs of the body,' she tells me. 'The same as: do you have a heart rate, are you breathing, what's your body temperature. It's a really strong indication of your health.' And yet 'it's also an area where there is so much taboo and misinformation.' As for family planning, Tin points out that 'there's been very little innovation since the pill came out in the 1950s. I mean, in the history of technology that's a *really* long time.'

Tin set up Clue because she wanted to 'enable women to take control of their own body and lives', but the motivation was also personal. She'd tried the pill, but, like many women, she'd had side effects. 'And I hadn't had any children so an IUD wasn't ideal. So I'd been using condoms for fifteen years.' In frustration, Tin started looking at patent databases, but '*everything* was about putting hormones into the body', she tells me. 'And I felt it was a very non-data-driven approach to this problem. It made me a little bit provoked, like: why is it that nobody has given this some serious effort and consideration? It's a pretty basic need for humankind.'

When she had the idea for the menstrual-tracking app there were only a couple of period-tracking apps available. 'And they were very first-generation products – basically a calendar that can count to twenty-eight. And if only our biology were that simple', she laughs. After a decade of being in the sector, Tin says, the science is still riddled with gaps. 'There really is a lack of data,' she tells me. Menstruation has been 'not just overlooked, but borderline actively ignored. We do a lot of work together with science institutions because there are really a lot of blank areas on the academic map. Like, what's even considered a normal bleeding pattern for an adolescent woman? That's one of the things we've been working on with Stanford. Science just doesn't know what's normal.'

Given the male domination of VCs, data gaps are perhaps particularly problematic when it comes to tech aimed at women. 'If you don't have good data,' explains Tin 'it's harder to open people's minds that something might be an issue if they don't encounter it themselves.' Boler agrees. 'We did talk to some VC investors who didn't believe [Elvie] was an interesting proposition,' she tells me.

The other problem women face when it comes to getting investment is 'pattern recognition'.[13] A corollary of 'culture fit', pattern recognition sounds data-driven, but it's basically just a fancy term for looks-similar-to-something-that-has-worked-in-the-past – where 'something' could be white-male-founder-who-dropped-out-of-Harvard-and-wears-hoodies. Genuinely: I dated a guy who was working on a start-up and he referenced this uniform when he was talking about getting funding. Hoody-based pattern recognition is real. And this emphasis on recognising a typically male pattern may be exacerbated by the common belief that tech is a field where inborn 'genius' (which, as we've seen, is stereotypically associated with men[14]) is more important than working hard (hence fetishising Harvard dropouts).

It all feels rather catch-22ish. In a field where women are at a disadvantage specifically *because* they are women (and therefore can't hope to fit a stereotypically male 'pattern'), data will be particularly crucial for female entrepreneurs. And yet it's the female entrepreneurs who are less likely to have it, because they are more likely to be trying to make products for women. For whom we lack data.

Still, some do manage to break through. Tin and Boler got their funding (Boler in part from Woskow). And now these specific data gaps are starting to be filled. Before they launched, Chiaro had over 150 women test their pelvic-floor trainer, Boler tells me. 'But we now have data on over a million workouts and we have a lot of measurements around pelvic-floor health which just haven't existed before.' This, she says, is the 'exciting thing about wearable tech: giving people better information about their bodies so they can make more informed decisions'.

*

But while Boler's and Tin's products may give women better information about their bodies, the same can't be said for all new tech, wearable or otherwise. In the tech world, the implicit assumption that men are the default human remains king. When Apple launched its health-monitoring system with much fanfare in 2014, it boasted a 'comprehensive' health tracker.[15] It could track blood pressure; steps taken; blood alcohol level; even molybdenum (nope, me neither) and copper intake. But as many women pointed out at the time, they forgot one crucial detail: a period tracker.[16]

This was not to be the only time Apple completely forgot about at least 50% of their users. When Apple launched their AI, Siri, she (ironically) could find prostitutes and Viagra suppliers, but not abortion providers.[17] Siri could help you if you'd had a heart attack, but if you told her you'd been raped, she replied 'I don't know what you mean by 'I was raped.'[18] These are basic errors that surely would have been caught by a team with enough women on it – that is, by a team without a gender data gap.

Products marketed as gender-neutral that are in fact biased towards men are rife across the (male-dominated) tech industry. From smartwatches that are too big for women's wrists,[19] to map apps that fail to account for women's desire for 'safest' in addition to 'fastest' routes; to 'measure how good you are at sex' apps called 'iThrust'[20] and 'iBang'[21] (and yes the in-built assumptions of what constitutes good sex are exactly what the names imply), the tech industry is rife with examples of tech that forget about women. Virtual reality (VR) headsets that are too big for the average woman's head; a 'haptic jacket' (a jacket that simulates touch) that fits snugly on a male body, but on a female reviewer's body 'could have fit over a puffy winter coat'; augmented-reality glasses whose lenses are too far apart for a woman to focus on the image, 'or whose frames immediately fall off my face'. Or, as I know from my experience of going on TV and giving public lectures, mic packs that require either a waistband or substantial pockets to attach to. Out goes pretty much every dress ever designed.

Defaulting to male seems particularly endemic in sports tech. Starting with the most basic, the calorie count on treadmills is perfect for practically no one, but it will be more accurate for your average man because its calculations are based on the average male weight (the default setting for calorie count on most exercise machines is for a person who weighs eleven stone). And although you can change the weight setting, that still leaves a calculation based on an average male calorie burn. Women generally have a higher fat and lower muscle distribution than men as well as different ratios of various muscle fibres. What this means at a basic level is that even after accounting for weight difference, men on average will burn 8% more calories than a woman of the same weight. The treadmill does not account for this.

There's no reason to think that things improved much with the advent of wearables, either. One study of twelve of the most common fitness monitors found that these underestimated steps during housework by up to 74% (that was the Omron, which was within 1% for normal walking or running) and underestimated calories burned during housework by as much as 34%.[22] Anecdotally, Fitbits apparently fail to account for movement while doing the extremely common female activity of pushing a pram (yes of course men push prams too, but not as often as the women who do 75% of the world's unpaid care). Another study, which unusually did manage to include almost 50% female participants, found that fitness devices were overestimating calorie burn by significant amounts.[23] Unfortunately, they failed to disaggregate their data so it is impossible to know if there were any sex differences.

Tech developers even forget women when they form the potential majority of customers. In the US, women make up 59% of people over the age of sixty-five and 76% of those living alone, suggesting a potential greater need for assistive technology like fall-detection devices.[24] The data we have suggests that not only do older women fall more often than men, they also injure themselves more when they do.[25] Data analysis of a month's worth of emergency department visits in the US found that of the 22,560 patients seen for fall injuries, 71%, were women. The rate of fracture was 2.2 times higher in women, and women had a hospitalisation rate 1.8 times that of men.[26]

And yet despite women's arguably greater need (as well as research indicating that women tend to fall differently, for different reasons, and in different places), gender analysis is missing from the development of this technology. In one meta-analysis of fifty-three fall detection device studies, only half of them even described the sex of participants, let alone delivered sex-disaggregated data;[27] another study noted that 'Despite extensive literature on falls among seniors, little is known about gender-specific risk factors.'[28]

The Proceedings of the 2016 International Conference on Intelligent Data Engineering and Automated Learning points out that 'a notable motivation for elders to reject fall-detection devices is their size', suggesting mobile phones as a solution.[29] Except this isn't really a solution for women because as the authors themselves note, women tend to keep their phones in their handbags, 'where fall-detection algorithms will likely fail because they are trained to detect falls through acceleration sensors close to the body trunk'.

In acknowledging this, the authors are unusual. Whitney Erin Boesel, a researcher at the Berkman Center for Internet and Society at Harvard, is a member of the 'quantified self' community, which promises 'self-knowledge through numbers'. These numbers are often collected via passive tracking apps on your phone, the classic being how many steps you've taken that day. But there's a pocket-sized problem with this promise: 'Inevitably some dude gets up at a conference and [says] something about how your phone is always on you,' Boesel told the *Atlantic*.[30] 'And every time I'll stand up, and I'll be like, "Hi, about this phone that is always on you. This is my phone. And these are my pants."'

Designing passive tracking apps as if women have pockets big enough to hold their phones is a perennial problem with an easy solution: include proper pockets in women's clothing (she types, furiously, having just had her phone fall out of her pocket and smash on the floor for the hundredth time). In the meantime, however, women use other solutions, and if tech developers don't realise women are being forced into workarounds, they may fail in their development.

A Cape Town-based tech company fell into this trap when they developed an app to help community health workers monitor HIV-positive patients. The app 'fulfilled all the usability requirements; it was easy to use, adaptable to local language' and solved a very specific issue. More than this, the community health workers were 'excited at the prospect of using it'.[31] But when the service was launched, it proved to be a flop. Despite several attempts to solve it, the problem remained a mystery until a new design team took over the project. A team that happened to have a woman in it. And this woman 'took only a day to discover the problem'. It turned out that in order to more safely complete their daily commute into the townships where their patients lived, female health workers were concealing their valuables in their underwear. And the phone was too big to fit in their bras.

Gender affects the kinds of questions we ask, says Margaret Mitchell, senior research scientist at Google. Limiting AI developers to one gender, she told Bloomberg News, puts companies 'in a position of myopia'.[32] Gayna Williams, ex-director of user experience at Microsoft, agrees.[33] In a blogpost titled 'Are you sure your software is gender-neutral?' Williams explains that all product design begins by deciding which problem needs solving. And that is all a matter of perception: what problem were NASA scientists solving when they decided to give Valkyrie, their space-navigation robot, breasts?[34]

On the topic of sexy robots, even if men do identify a problem that affects us all, that doesn't mean that without female input they will come up with the right solution. When, in 'retaliation' for women denying him the sex to which he believed he was entitled, Alek Minassian mowed down and killed ten people in Toronto with a rented van, the *New York Times* published a column headlined 'The Redistribution of Sex', which argued that sex robots could be the answer to the plight of men who can't convince women to sleep with them. Feminists might argue that the solution is, instead, to challenge male sexual entitlement.

When it comes to the tech that ends up in our pockets (I'm ever hopeful), it all comes down to who is making the decisions. And like the world of venture capitalists, the tech industry is dominated by men. Margaret Mitchell calls this the 'sea of dudes' problem.[35] Over the past five years she's worked with around ten women and 'hundreds' of men. Across 'professional computing' as a whole in the US, 26% of jobs are held by women compared to the 57% of jobs women hold across the entire US workforce.[36] In the UK, women make up 14% of the STEM workforce.[37]

As well as a rash of sexy robots, the sea of dudes leads to products like the 'enormous robot research prototype called PR2' that computer scientist and co-founder of a robotics company Tessa Lau encountered when she worked for robotics research lab Willow Garage. It weighed 'hundreds of pounds – it's much larger than a smaller woman – and it has two big arms. It looks really scary. I didn't even want one of those things near me if it wasn't being controlled properly.' When I interviewed her a couple of years ago, roboticist Angelica Lim told me a similar story about the robot she encountered at a conference in Slovenia, which would come and shake your hand if you waved at it. When she waved at this 5'8' robot on wheels (the average American woman is 5'4') the robot slowly turned towards her, put out its hand, and then came 'barrelling towards me, fast', making her jump backwards and shriek.

Contrast these examples to the virtual-reality headset trialled by tech journalist Adi Robertson.[38] The headset was meant to track her eyes, but it didn't work for her – until an employee asked if she was wearing mascara. 'When it got recalibrated perfectly a few minutes later, I was surprised – not by the fact that it worked, but by the fact that anyone had thought to troubleshoot make-up. Incidentally,' she writes, 'this was one of the only VR start-ups I've ever covered with a female founder.'

Most VR companies aren't founded by women, however, and so the VR experience often comes with an in-built male bias. Like much of the online world, VR gaming seems to have a sexual harassment problem – and this problem is something VR's mainly male developers are routinely forgetting to account for.[39]

When author and gamer Jordan Belamire tried the VR game *QuiVr* in multiplayer mode, she was sexually assaulted by another user called BigBro442.[40] 'Virtual' makes it sound like it isn't real – but it felt real to Belamire. And no wonder. VR is meant to feel real, and it can be so successful at tricking your brain that it is being explored as a treatment for PTSD, phobias, even phantom-limb syndrome.[41]

To be fair to the male designers of *QuiVr*, they had an excellent and proactive response to Belamire's blog.[42] They immediately redesigned their 'Personal Bubble' setting (in which other player's hands disappear if they come close to your face) to cover the entire body and so make such groping impossible. But as they themselves noted, while they had thought 'of the possibility of some silly person trying to block your view with their hands and ruining the game', they hadn't thought of extending the fading function to the rest of the body. How, they asked, 'could we have overlooked something so obvious?'

Fairly simply, to be honest. Henry Jackson and Jonathan Schenker are clearly well-meaning men who don't intend to shut women out. But it's Sergey Brin and the pregnancy parking all over again: even the best of men can't know what it's like to go through the world as a person with a body which some other people treat as an access-all-areas amusement arcade. This just isn't something that Jackson and Schenker have to face on a regular basis, and therefore it really isn't all that surprising that they missed 'something so obvious'.

Male violence is far from the only issue keeping women out of VR. From the oversized headsets, to research showing that VR causes motion sickness in women to a far larger degree than it does in men,[43] to the fact that narrow computer displays favour men in tasks that require spatial awareness,[44] you're left with another platform that just doesn't work well for women – and that is therefore likely to have fewer women on it.

We don't know exactly why women are more likely to experience motion sickness while using VR, but Microsoft researcher danah boyd conducted a study that suggests a possible explanation.[45] Human eyes use two basic cues to determine depth: 'motion parallax' and 'shape-from-shading'. Motion parallax refers to how an object seems bigger or smaller depending on how close you are to it, while shape-from-shading refers to the way the shading of a point changes as you move. And while 3D VR is pretty good at rendering motion parallax, it still does 'a terrible job' of emulating shape-from-shading.

This discrepancy creates sex differences in how well VR works, because, as boyd discovered, men are 'significantly more likely' to rely on motion parallax for depth perception, while women rely on shape-from-shading. 3D environments are literally sending out information signals that benefit male over female depth perception. The question is: would we be so behind on recreating shape-from-shading if we had been testing 3D VR on equal numbers of men and women from the start?

Tom Stoffregen, professor of kinesiology at the University of Minnesota, has an entirely different theory for why women experience more motion sickness than men. The classic theories, he says, are 'focused almost entirely on sensory stimulation'. The idea is that what you feel in your inner ear doesn't match up with what you see with your eyes. And 'that's true,' says Stoffregen, 'but it's not the only thing that changes. The big thing that traditional theories have not talked about,' he says, 'is changes in what you need to do to control your body.'

In the normal course of your day, your body is constantly making micro-adjustments in order to keep you stable. When you stand, when you sit, when you walk. But when you're in a moving environment – say a car, or a ship – what you have to do to remain stable changes because your body is being destabilised. So, says Stoffregen, 'You are being physically required to move in a different way and you haven't learnt how to do it yet.' And like cars and ships, he says, VR destabilises the body. Hence the motion sickness.

The VR industry has so far shown little interest in Stoffregen's research. 'They understand that it's a serious problem,' but they're going the wrong way about fixing it, he says. 'The people who design VR think it's just an object that you put in front of your eyes, and the idea that it would have anything to do with anything *other* than your eyes is incoherent to them.' But, he says, VR developers have to understand that they're doing more than 'simply placing screens in front of people's eyes. Whether they like it or not and whether they *know* it or not.'

VR developers also have to start collecting data systematically – and separating it by sex. 'Most of the data on motion sickness in VR are anecdotal,' explains Stoffregen, 'and they come from the people who are working in these companies just using the systems themselves or trying them out at computer technology conferences or whatever. So they're completely unsystematic – and most of these people are men.'

One of the most convincing aspects of Stoffregen's theory is how it finally explains why I get car sick in every seat other than the driving seat: it's all about control. When you're walking, you are in control of your movements. You know what's coming. On a ship, or in a car, someone else is in control – unless you're the driver. 'The driver knows what the motion of the car is going to be and so the driver is able to stabilise his or herself in what we call an anticipatory fashion,' explains Stoffregen, 'whereas the passenger cannot know in quantitative detail what the car is going to be doing. And so their control of their own body must be compensatory. And anticipatory control is just better than compensatory control. You know, that ain't no rocket science.'

But where does the sex difference come in? 'Everybody who studies motion sickness has known that women are more susceptible than men basically forever,' says Stoffregen. 'It's just an utterly uncontroversial fact. It's just *there*.' But, and he includes himself in this group, 'very few people did any research on it or made any effort to try and figure that out.' *Plus ça change.*

But in 2010, Stoffregen made a discovery. 'I was sort of fooling around in the literature and I came across some results that I didn't know about,' and which showed that there are sex differences in body sway. 'These are small subtle differences. You can't just watch somebody and see them, but in terms of the subtle quantitative details of how the body moves back and forth, there are in fact reliable sex differences. And as soon as I saw this, I mean, I'm talking the *minute* that I saw this I knew that oh, OK, now I have something to say about the sex difference in motion sickness. Because my whole story about motion sickness is that it's related to the control of the body.' Since then, Stoffregen has also discovered evidence that 'women's postural sway changes across the menstrual cycle'. And this is significant because 'a woman's susceptibility to motion sickness changes across the menstrual cycle. And those two things link up, believe it or not.'

A considerable gender data gap remains. We don't yet know exactly how and when women's body-sway changes. But as a woman who suffers from extreme car sickness, I am excited and enraged by Stoffregen's findings, particularly because of how it ties into another gender data gap I've been looking into: car design.

When you're sitting down, you're still swaying. 'If you're sitting on a stool then you're swaying around your hips,' explains Stoffregen. 'If your chair has a back, then your head is swaying on your neck. The only way to really get rid of that is to have a headrest and to use it,' he adds. And I feel like one of those cartoon light bulbs has just gone off in my head. What if the headrest is at the wrong height, at the wrong angle, and the wrong shape to accommodate your body? Could women's increased propensity to motion sickness in cars be exacerbated by cars being designed around the male body, I ask. 'I think that's quite possible, sure,' Stoffregen replies. 'The quality of the stabilisation, if it's the wrong height or whatever … what you're telling me is new to me, but it sounds perfectly plausible.'

But here I run into yet another data gap: the available research on whether car headrests have been designed to account for the female body is seemingly non-existent. This gap is hardly unexpected though: car design has a long and ignominious history of ignoring women.

Men are more likely than women to be involved in a car crash, which means they dominate the numbers of those seriously injured in car accidents. But when a woman is involved in a car crash, she is 47% more likely to be seriously injured than a man, and 71% more likely to be moderately injured,[46] even when researchers control for factors such as height, weight, seat-belt usage, and crash intensity.[47] She is also 17% more likely to die.[48] And it's all to do with how the car is designed – and for whom.

Women tend to sit further forward than men when driving. This is because we are on average shorter. Our legs need to be closer to reach the pedals, and we need to sit more upright to see clearly over the dashboard.[49] This is not, however, the 'standard seating position'. Women are 'out of position' drivers.[50] And our wilful deviation from the norm means that we are at greater risk of internal injury on frontal collisions.[51] The angle of our knees and hips as our shorter legs reach for the pedals also makes our legs more vulnerable.[52] Essentially, we're doing it all wrong.

Women are also at higher risk in rear-end collisions. Women have less muscle on our necks and upper torso than men, which make us more vulnerable to whiplash (by up to three times[53]), and car design has amplified this vulnerability. Swedish research has shown that modern seats are too firm to protect women against whiplash injuries: the seats throw women forward faster than men because the back of the seat doesn't give way for women's on average lighter bodies.[54] The reason this has been allowed to happen is very simple: cars have been designed using car-crash test dummies based on the 'average' male.

Crash-test dummies were first introduced in the 1950s, and for decades they were based around the fiftieth percentile male. The most commonly used dummy is 1.77 m tall and weighs 76 kg (significantly taller and heavier than an average woman), and the dummy also has male muscle-mass proportions and a male spinal column. In the early 1980s, researchers argued for the inclusion of a fiftieth percentile female in regulatory tests, but this advice was ignored.[55] It wasn't until 2011 that the US started using a female crash-test dummy,[56] although, as we'll see, just how 'female' these dummies are is questionable.

In 2018, Astrid Linder, research director of traffic safety at the Swedish National Road and Transport Research Institute, presented a paper at the Road Safety on Five Continents Conference in South Korea in which she ran through EU regulatory crash-test requirements.[57] In the EU, there are five tests a car must pass before being allowed on the market: one safety-belt test, two frontal-collision tests, and two lateral-collision tests. In no test is an anthropometrically correct female crash-test dummy required. The seat-belt test, one of the frontal-collision tests, and both lateral-collision tests all specify that a fiftieth-percentile male dummy should be used. When Linder looked at regulatory tests worldwide, she found that while there are 'several local differences', regulatory tests are still using the fiftieth-percentile male 'to represent the whole adult population'.

There is one EU regulatory test that requires what is called a fifth-percentile female dummy, which is meant to represent the female population. Only 5% of women will be shorter than this dummy. But there are a number of data gaps. For a start, this dummy is only tested in the passenger seat, so we have no data at all for how a female driver would be affected – something of an issue you would think, given women's 'out of position' driving position. And secondly, this female dummy is not really female. It is just a scaled-down male dummy.

Consumer tests can be slightly more stringent. When I spoke to the EuroNCAP (a Eurpean organisation that provides car safety ratings for consumers) they informed me that since 2015 they have used male and female dummies in both front-crash tests and that they do base their female dummies on female anthropometric data – with the caveat that this is 'where data is available'. And that, says Linder, is quite the caveat: 'To my knowledge, little or even no such data' has been applied to crash dummies. In any case, EuroNCAP acknowledged that 'sometimes' they do just use scaled-down male dummies. But, as we'll cover extensively in the next chapter, women are not scaled-down men. We have different muscle-mass distribution. We have lower bone density. There are sex differences in vertebrae spacing. As Stoffregen has noted, even our body sway is different. And these differences are all crucial when it comes to injury rates in car crashes.

The situation is even worse for pregnant women. Although a pregnant crash-test dummy was created back in 1996, testing with it is still not government-mandated either in the US or in the EU.[58] In fact, even though car crashes are the number-one cause of foetal death related to maternal trauma,[59] we haven't even yet developed a seat belt that works for pregnant women. Research from 2004 suggests that pregnant women should use the standard seat belt,[60] but 62% of third-trimester pregnant women don't fit the standard seat-belt design.[61] A three-point seat belt can also ride up on women who carry low, which a 1996 study found can treble or quadruple force transmission to the abdomen compared to when the belt is worn below the uterus, 'with a corresponding increased risk of fetal injury'.[62] Standard seat belts aren't great for non-pregnant women either: apparently, in an effort to accommodate our breasts many of us are wearing seat belts 'improperly' which again, increases our risk of injury (another reason we should be designing explicitly female dummies rather than just smaller male dummies).[63] And it's not just a woman's belly that changes in pregnancy: breast-size changes can also diminish seat-belt efficacy by affecting positioning. Here again, we find an example of a situation where we have the data on women, but are just ignoring it. Clearly what is needed is a wholesale redesign of cars using complete data, and this should be fairly simple since it's not exactly hard to find women to model a test dummy on.

Even with all these gaps, the 2011 introduction of the female crash-test dummy in the US still sent cars' star ratings plummeting. The *Washington Post* reported on the experience of Beth Milito and her husband, who bought a 2011 Toyota Sienna, based primarily on its four-star safety rating.[64] But all was not as it seemed. The passenger seat, which Milito says she is likely to be sitting in when they are 'out and about as a family', had a two-star rating. In the previous year's model, the front passenger seat (tested on a male dummy) had earned a top five-star rating. But the shift to female dummies revealed that in a front collision at 35 mph a female passenger had a 20–40% risk of being killed or seriously injured. The average risk of death for that class of vehicle, explains the *Washington Post*, is 15%.

A 2015 report by the Insurance Institute for Highway Safety is excitingly headlined 'Improved vehicle designs bring down death rates' – which sounds great. Perhaps this is the result of the new legislation? Unlikely. Buried in the report is the following telltale line: 'The rates include only driver deaths because the presence of passengers is unknown.' This is a huge gender data gap. When men and women are in a car together, the man is most likely to be driving.[65] So not collecting data on passengers more or less translates as not collecting data on women.

The infuriating irony of all this is that the gendered passenger/driver norm is so prevalent that, as we've seen, the passenger seat is the only seat that is commonly tested with a female crash-test dummy anyway, with the male crash-test dummy still being the standard dummy for the driver's seat. So stats that include only driver fatalities tell us precisely zero about the impact of introducing the female crash-test dummy. In conclusion, a more accurate headline for the report would be 'Improved vehicle design brings down death rates in the seat most likely to be occupied by men, but who knows about death rates in the seat most likely to be occupied by women even though we already know women are 17% more likely to die in a car crash.' Admittedly, this is less snappy.

When I speak to Dr David Lawrence, director of safety-literature database the SafetyLit Foundation, he tells me that 'in most US states the quality of police crash reports is at best poor for use as a research tool'. Little data is gathered about anyone other than the driver. Written police reports have often been handed to 'contract companies for data entry', most of whom use prison labour for data entry. 'Data-quality checks were rare and when quality was evaluated it was found wanting. For example, in Louisiana for most crashes in the 1980s most of the occupants were males who were born on January 1st, 1950. Almost all of the vehicles involved in crashes were the 1960 model year.' Except they weren't. These were just the default settings.

Lawrence tells me that even though this problem has been found in 'many other states', the data hasn't improved 'because no changes in the data-entry practices were made. The federal government required that states provide police crash-report data to NHTSA (The National Highway Traffic Safety Administration) but set no standards for data quality nor penalty for sending junk data.'

Astrid Linder has been working on what she says will be the first crash-test dummy to accurately represent female bodies. Currently, it's just a prototype, but she is calling on the EU to make testing on anthropometrically correct female crash-test dummies a legal requirement. In fact, Linder argues that this technically already *is* a legal requirement. Article 8 of the legally binding Treaty of the Functioning of the European Union reads, 'In all its activities, the Union shall aim to eliminate inequalities, and to promote equality, between men and women.'[66] Clearly, women being 47% more likely to be seriously injured in a car crash is one hell of an inequality to be overlooking.

In some ways it's hard to understand why a proper female crash-test dummy hasn't been developed and made a legal requirement in car tests years ago. But on the other hand, and given all we know about how women and their bodies are routinely ignored in design and planning, it's not surprising at all. From development initiatives to smartphones, from medical tech to stoves, tools (whether physical or financial) are developed without reference to women's needs, and, as a result these tools are failing them on a grand scale. And this failure affects women's lives on a similarly grand scale: it makes them poorer, it makes them sicker, and, when it comes to cars, it is killing them. Designers may believe they are making products for everyone, but in reality they are mainly making them for men. It's time to start designing women in.

PART IV

Going to the Doctor

CHAPTER 10

The Drugs Don't Work

It took twelve years for Michelle to receive a diagnosis. 'I was about fourteen when I first started having symptoms,' she tells me. 'I was too ashamed to go to a doctor for it.' She kept her urgent, painful, frequent, sometimes bloody bowel movements a secret for two years, until one night, it hurt too much to hide anymore. 'I couldn't move from the foetal position on my bathroom floor. I was afraid I was dying.' She was sixteen.

Michelle's parents rushed her to the emergency room. A doctor there asked her (in front of her parents) if she could be pregnant. No, she couldn't be, Michelle explained, because she hadn't had sex, and in any case, the pain was in her intestines. 'They wheeled me into an exam room and without any explanation, placed my feet into stirrups. The next thing I knew, a large, cold metal speculum was crammed in my vagina. It hurt so badly I sat up and screamed and the nurse had to push me back down and hold me there while the doctor confirmed that indeed, I was not pregnant.' She was discharged with 'nothing more than some overpriced aspirin and the advice to rest for a day'.

Over the next decade Michelle sought help from two more doctors and two (male) gastroenterologists, both of whom told her that her problems were in her head and that she needed to be less anxious and stressed. At the age of twenty-six Michelle was referred to a female GP who scheduled her for a colonoscopy: it revealed that the entire left side of her colon was diseased. She was diagnosed with both irritable bowel syndrome and ulcerative colitis. 'Funnily enough', Michelle says, 'my colon is not in my head.' As a result of the extended delay in receiving a diagnosis and treatment she has been left with an increased risk of colon cancer.

It's hard to read an account like this and not feel angry with the doctors who let Michelle down so badly. But the truth is that these are not isolated rogue doctors, bad apples who should be struck off. They are the products of a medical system which, from root to tip, is systematically discriminating against women, leaving them chronically misunderstood, mistreated and misdiagnosed.

It begins with how doctors are trained. Historically it's been assumed that there wasn't anything fundamentally different between male and female bodies other than size and reproductive function, and so for years medical education has been focused on a male 'norm', with everything that falls outside that designated 'atypical' or even 'abnormal'.[1] References to the 'typical 70 kg man'[2] abound, as if he covers both sexes (as one doctor pointed out to me, he doesn't even represent men very well). When women *are* mentioned, they are presented as if they are a variation on standard humanity. Students learn about physiology, and female physiology. Anatomy, and female anatomy. 'The male body', concluded social psychologist Carol Tavris in her 1992 book *The Mismeasure of Woman*, 'is anatomy itself.'[3]

This male-default bias goes back at least to the ancient Greeks, who kicked off the trend of seeing the female body as a 'mutilated male' body (thanks, Aristotle). The female was the male 'turned outside in'. Ovaries were female testicles (they were not given their own name until the seventeenth century) and the uterus was the female scrotum. The reason they were inside the body rather than dropped out (as in typical humans) is because of a female deficiency in 'vital heat'. The male body was an ideal women failed to live up to.

Modern doctors of course no longer refer to women as mutilated males, but the representation of the male body as *the* human body persists. A 2008 analysis of a range of textbooks recommended by twenty of the 'most prestigious universities in Europe, the United States and Canada' revealed that across 16,329 images, male bodies were used three times as often as female bodies to illustrate 'neutral body parts'.[4] A 2008 study of textbooks recommended by Dutch medical schools found that sex-specific information was absent even in sections on topics where sex differences have long been established (such as depression and the effects of alcohol on the body), and results from clinical trials were presented as valid for men and women even when women were excluded from the study.[5] The few sex differences that did get a mention were 'hardly accessible via index or layout', and in any case tended to be vague one-liners such as 'women, who more often have atypical chest discomfort'. (As we'll see, only one in eight women who have a heart attack report the classic male symptom of chest pain, so in fact this description is arguably not only vague, but inaccurate.[6])

In 2017 I decided to see if much had changed, and set off to a large bookshop in central London with a particularly impressive medical section. Things had not changed. The covers of books entitled 'Human Anatomy' were still adorned with be-muscled men. Drawings of features common to both sexes continued to routinely include pointless penises. I found posters entitled 'Ear, Nose & Throat', 'The Nervous System', 'The Muscular System', and 'The Vascular System and Viscera', all of which featured a large-scale drawing of a man. The vascular-system poster did, however, include a small 'female pelvis' off to one side, and me and my female pelvis were grateful for small mercies.

The gender data gaps found in medical textbooks are also present in your typical medical-school curriculum. A 2005 Dutch study found that sex- and gender-related issues were 'not systematically addressed in curriculum development'.[7] A 2006 review of 'CurrMIT', the US online database for med-school courses, found that only nine out of the ninety-five schools that entered data into the system offered a course that could be described as a 'women's health course'.[8] Only two of these courses (obstetrics and gynaecology classes taught in the second or third academic years) were mandatory. Even conditions that are known to cause the greatest morbidity and mortality in women failed to incorporate sex-specific information. Ten years later, another review found that the integration of sex- and gender-based medicine in US med schools remained 'minimal' and 'haphazard', with gaps particularly identified in the approach to the treatment of disease and use of drugs.[9]

These gaps matter because contrary to what we've assumed for millennia, sex differences can be substantial. Researchers have found sex differences in every tissue and organ system in the human body,[10] as well as in the 'prevalence, course and severity' of the majority of common human diseases.[11] There are sex differences in the fundamental mechanical workings of the heart.[12] There are sex differences in lung capacity,[13] even when these values are normalised to height (perhaps related is the fact that among men and women who smoke the same number of cigarettes, women are 20–70% more likely to develop lung cancer[14]).

Autoimmune diseases affect about 8% of the population,[15] but women are three times more likely to develop one, making up about 80% of those affected.[16] We don't fully know why, but researchers think it might be down to women being the child-bearing sex:

the theory is that females 'evolved a particularly fast and strong immune response to protect developing fetuses and newborn babies',[17] meaning that sometimes it overreacts and attacks the body.[18] The immune system is also thought to be behind sex-specific responses to vaccines: women develop higher antibody responses and have more frequent and severe adverse reactions to vaccines,[19] and a 2014 paper proposed developing male and female versions of influenza vaccines.[20]

Sex differences appear even in our cells: in blood-serum biomarkers for autism;[21] in proteins;[22] in immune cells used to convey pain signals;[23] in how cells die following a stroke.[24] A recent study also found a significant sex difference in the 'expression of a gene found to be important for drug metabolism'.[25] Sex differences in the presentation and outcome of Parkinson's disease, stroke and brain ischaemia (insufficient blood flow to the brain) have also been tracked all the way to our cells,[26] and there is growing evidence of a sex difference in the ageing of the blood vessels, 'with inevitable implications for health problems, examination and treatment'.[27] In a 2013 *Nature* article, Dr Elizabeth Pollitzer points to research showing that male and female mice cells have been found to respond differently to stress; that male and female human cells 'exhibit wildly different concentrations of many metabolites'; and to 'mounting evidence' that 'cells differ according to sex irrespective of their history of exposure to sex hormones'.[28]

There are still vast medical gender data gaps to be filled in, but the past twenty years have demonstrably proven that women are not just smaller men: male and female bodies differ down to a cellular level. So why aren't we teaching this?

The inclusion of sex-specific information in textbooks is dependent on the availability of sex-specific data, but because women have largely been excluded from medical research this data is severely lacking. Even the very basics of sex determination have a sex data gap: since the landmark 1990 paper that identified the Y chromosome as 'the' sex-determining region, the female sex has – the irony – been seen as the default. But in this case, the default didn't mean we focused on the female. Rather, research instead focused on testes development as the supposedly 'active' process, while female sexual development was seen as a passive process – until 2010, when we finally started researching the active process of ovarian determination.[29]

Most early research into cardiovascular disease was conducted on men, and women continue to be under-represented, making up only 25% of participants across thirty-one landmark trials for congestive heart failure between 1987 and 2012.[30] Women represent 55% of HIV-positive adults in the developing world,[31] and in parts of Africa and the Caribbean women aged five to twenty-four are up to six times more likely to be HIV-positive than young men of the same age.[32] We also know that women experience different clinical symptoms and complications due to HIV, and yet a 2016 review of the inclusion of women in US HIV research found that women made up only 19.2% of participants in antiretroviral studies, 38.1% in vaccination studies and 11.1% in studies to find a cure.[33]

Because of their routine exclusion from clinical trials we lack solid data on how to treat pregnant women for pretty much anything. We may not know how a disease will take hold or what the likely outcome may be, although the WHO warns that many diseases can have 'particularly serious consequences for pregnant women, or can harm the foetus'.[34] Some strains of influenza virus (including the 2009 H1N1 swine flu virus) have 'particularly severe symptoms during pregnancy'. There is also evidence that SARS can be more severe during pregnancy. It is of course understandable that a pregnant woman may be reluctant to take part in medical research, but this doesn't mean that we have to just throw our hands up in the air and accept that we know nothing: we should be routinely and systematically tracking, recording and collating pregnant-women's health outcomes. But we aren't – not even during pandemics: during the 2002–4 SARS outbreak in China, pregnant-women's health outcomes were not systemically tracked and 'consequently', the WHO points out, 'it was not possible to fully characterize the course and outcome of SARS during pregnancy'.[35] Another gender data gap that could have been so easily avoided, and information that will be lacking for when the next pandemic hits.

Like the failure to include women in anatomy textbooks, the failure to include women in medical trials is a historical problem that has its roots in seeing the male body as the default human body, but this traditional bias was radically enhanced in the 1970s, to the great detriment of women's health, following one of the biggest medical scandals of the twentieth century.[36]

In 1960 doctors began prescribing thalidomide to pregnant women who suffered from morning sickness. The drug, which had been available as a mild over-the-counter sedative in many countries since the late 1950s, was considered safe because its developers 'could not find a dose high enough to kill a rat'.[37] But while it didn't kill rats, it did affect foetal development (something that in fact the manufacturers knew as early as 1959).[38] Before the drug was taken off the market in 1962, over 10,000 children had been born around the world with thalidomide-related disabilities.[39] In the wake of the scandal, the US Food and Drug Administration (FDA) issued guidelines in 1977 excluding women of childbearing potential from drug trials. This exclusion went unquestioned.[40] The acceptance of the male norm went unquestioned.

The male norm continues to go unquestioned by many today, with some researchers continuing to insist, in the face of all the evidence, that biological sex doesn't matter. One public-health researcher revealed that she had received the following feedback on two different grant applications: 'I wish you'd stop with all this sex stuff and get back to science', and 'I've been in this field for 20 years and this [biological difference] doesn't matter'.[41] It isn't just anonymous notes, either. A 2014 op-ed published in the journal *Scientific American* complained that including both sexes in experiments was a waste of resources;[42] in 2015 an op-ed in the official scientific journal of the US National Academy of Sciences insisted that 'focusing on preclinical sex differences will not address women's and men's health disparities'.[43]

Alongside insisting that sex differences don't matter, some researchers advocate against the inclusion of women in research on the basis that while biological sex may matter, the lack of comparable data arising from the historical data gap makes including women inadvisable (talk about adding insult to injury).[44] Female bodies (both the human and animal variety) are, it is argued, too complex, too variable,[45] too costly to be tested on. Integrating sex and gender into research is seen as 'burdensome'.[46] It is seen as possible for there to be 'too much gender',[47] and for its exclusion to be acceptable on the basis of 'simplification'[48] – in which case it's worth noting that recent studies on mice have actually shown greater variability in males on a number of markers.[49] So who's too complicated now?

Beyond the argument that women's bodies, with their fluctuating, 'atypical' hormones, are simply inconvenient research vessels, researchers also defend their failure to include women in trials by claiming that women are harder to recruit. And it is certainly true that, due to women's care-giving responsibilities they have less leisure time and may find it harder to make, for example, clinic appointments during the school run. However, this is an argument for adapting trial schedules to women, rather than simply excluding them, and in any case, it is possible to find women if you really want to. While reviews of FDA-mandated medical product trials found that women made up only 18% of participants in trials for endovascular occlusion devices (used if your foetal blood vessel hasn't closed of its own accord)[50] and 32% of participants in studies on coronary stents (which, incidentally, are another device where women have worse outcomes than men),[51] women represented 90% and 92% of participants in facial wrinkle correction trials and dental device trials, respectively.

A more novel approach to addressing the problem of female under-representation in medical research is simply to claim that there is no problem, and women are represented just fine, thank you very much. In February 2018 a paper was published in the *British Journal of Pharmacology* entitled 'Gender differences in clinical registration trials: is there a real problem?'[52] Following 'cross-sectional, structured research into publicly available registration dossiers of Food and Drug Administration (FDA)-approved drugs that are prescribed frequently', the all-male-authored paper concluded that, no, the problem was not 'real'.

Leaving aside any philosophical debate over what an unreal problem might be, the authors' conclusions are baffling. For a start, data was available for only 28% of the drug trials, so we have no way of knowing how representative the sample is. In the data researchers *were* able to access, the number of female participants in over a quarter of trials did not match the proportion of women in the US affected by the disease the drug was supposed to treat. Furthermore, the study did not address trials for generic drugs, which represent 80% of prescriptions in the United States.[53] The FDA describes a generic drug as 'a medication created to be the same as an already marketed brand-name drug' and they are sold after the patent for the original branded drug runs out. Drugs trials for generic drugs are much less rigorous than original trials, having only to demonstrate equal bioavailability, and they are conducted 'almost exclusively' in young adult males.[54] This matters because even with the same active ingredient, different inactive ingredients and different fabrication technology can affect a drug's potency.[55] And sure enough, in 2002 the FDA's Center for Drug Evaluation and Research showed 'statistically significant differences between men and women in bioequivalence for most generic drugs compared with reference drugs'.[56]

Despite all this, the authors claimed that there was no evidence of any systematic under-representation of women in clinical trials because in phase two and three trials women were included at 48% and 49%, respectively. But the study authors themselves report that in phase one trials women represented only 22% of participants. And, contrary to what their conclusion might imply, the under-representation of women in phase one trials does matter. According to the FDA, the second most common adverse drug reaction in women is that the drug simply doesn't work, even though it clearly works in men. So with that substantial sex difference in mind: how many drugs that *would* work for women are we ruling out at phase one trials just because they don't work in men?

Digging deeper into the numbers, another issue the authors completely failed to address is whether or not the drugs were tested in women at different stages in their menstrual cycles. The likelihood is that they weren't, because most drugs aren't. When women are included in trials at all, they tend to be tested in the early follicular phase of their menstrual cycle, when hormone levels are at their lowest – i.e. when they are superficially most like men. The idea is to 'minimise the possible impacts oestradiol and progesterone may have on the study outcomes'.[57] But real life isn't a study and in real life those pesky hormones will be having an impact on outcomes. So far, menstrual-cycle impacts have been found for antipsychotics, antihistamines and antibiotic treatments as well as heart medication.[58] Some antidepressants have been found to affect women differently at different times of their cycle, meaning that dosage may be too high at some points and too low at others.[59] Women are also more likely to experience drug-induced heart-rhythm abnormalities[60] and the risk is highest during the first half of a woman's cycle.[61] This can, of course, be fatal.

Finally, the authors didn't consider the number of drug treatments that might be beneficial to women but never even reach human testing because they were ruled out at the cell and animal trial stage. And this number could be substantial. Sex differences in animals have been consistently reported for nearly fifty years, and yet a 2007 paper found that 90% of pharmacological articles described male-only studies.[62] In 2014 another paper found that 22% of animal studies did not specify sex, and of those that did, 80% included only males.[63]

Perhaps most galling from a gender-data-gap perspective was the finding that females aren't even included in animal studies on female-prevalent diseases. Women are 70% more likely to suffer depression than men, for instance, but animal studies on brain disorders are five times as likely to be done on male animals.[64] A 2014 paper found that of studies on female-prevalent diseases that specified sex (44%), only 12% studied female animals.[65] Even when both sexes are included there is no guarantee the data will be sex-analysed: one paper reported that in studies where two sexes were included, two-thirds of the time the results were not analysed by sex.[66] Does this matter? Well, in the 2007 analysis of animal studies, of the few studies that did involve rats or mice of both sexes, 54% revealed sex-dependent drug effects.[67]

These sex-dependent effects can be extreme. Dr Tami Martino researches the impact of circadian rhythms on heart disease, and during a 2016 lecture to the Physiology Society she recounted a recent shock discovery. Together with her team, she conducted a study which found that the time of day you have a heart attack affects your chances of survival. A heart attack that hits during the day triggers, among other things, a greater immune response. In particular, it triggers a greater neutrophil response (neutrophils are a type of white blood cell that are usually first on the scene in response to any injury), and this response correlates with a better chance of survival. This finding has been replicated many times over many years with many different animals, becoming, explained Martino, the 'gold standard for survivorship in the literature'.

So Martino and her team were 'quite surprised' when in 2016 another group of researchers released a paper which also found that daytime heart attacks triggered a greater neutrophil response – but that this correlated with a *worse* chance of survival. After a substantial amount of head-scratching, they realised there was one basic difference between the historic studies and this one new study: the old studies had all used male mice, while this new paper had used female mice. Different sex: totally opposite result.

As for cell studies, a 2011 review of ten cardiovascular journals found that when sex was specified 69% of cell studies reported using only male cells.[68] And 'when sex was specified' is an important caveat: a 2007 analysis of 645 cardiovascular clinical trials (all published in prominent journals) found that only 24% provided sex-specific results.[69] A 2014 analysis of five leading surgical journals found that 76% of cell studies did not specify sex and of those that did, 71% included only male cells and only 7% reported sex-based results.[70] And again, even for diseases that are more prevalent in women, researchers can be found 'exclusively' studying XY cells.[71]

As in animal and human studies, when sex *has* been analysed in cell studies, dramatic differences have been found. For years researchers were puzzled by the unpredictability of transplanted muscle-derived stem cells (sometimes they regenerated diseased muscle, sometimes they didn't do anything) until they realised that the cells weren't unpredictable at all – it's just that female cells promote regeneration and male cells don't. Perhaps of more urgent concern for women's health is the 2016 discovery of a sex difference in how male and female cells respond to oestrogen. When researchers[72] exposed male and female cells to this hormone and then infected them with a virus, only the female cells responded to the oestrogen and fought off the virus. It's a tantalising finding that inevitably leads to the following question: how many treatments have women missed out on because they had no effect on the male cells on which they were exclusively tested?

In light of all this evidence, it's hard to see how researchers can continue to argue in good faith that sex doesn't matter. Rather, it seems clear that McGill University neuroscientist Jeffrey Mogil was right when he told the Organisation for the Study of Sex Differences that failing to include both sexes 'right at the very beginning' of your research 'is not only scientifically idiotic and a waste of money, it is an ethical issue as well'.[73] Nevertheless, women continue to be routinely under-represented in medical research, and you can't even expect sex-specific trials to adequately represent women. When the 'female Viagra'[74] that was released with much fanfare in 2015 was found to potentially interact negatively with alcohol (as most readers will know, the absorption of alcohol differs between men and women[75]), its manufacturer, Sprout Pharmaceuticals, quite rightly decided to run a trial – for which they recruited twenty-three men and two women.[76] They did not sex-disaggregate the data.

In this latter failure, they are not alone. Several reviews of papers published in major journals over the past ten years have all identified a routine failure to either present results by sex, or to explain why the influence of sex has been ignored.[77] A 2001 US Government Accounting Office (GAO) audit of FDA records found that about a third of documents didn't sex-disaggregate their outcomes and 40% didn't even specify the sex of the participants.

The auditors concluded that the FDA had 'not effectively overseen the presentation and analysis of data related to sex differences in drug development',[78] a finding that was confirmed in a 2007 analysis of new drug applications submitted to the FDA which found a failure to establish standards for data analysis of applications.[79] In 2015 the GAO criticised the US National Institutes of Health (NIH) for failing to routinely track whether researchers had actually evaluated any differences between the sexes.[80] Things are often even worse in non-government-funded trials – which represent the majority of studies. A 2014 investigation into sex analysis in cardiovascular trials found that thirty-one of sixty-one NIH-sponsored trials analysed outcomes by sex compared with only 125 of 567 non-NIH-sponsored clinical trials.[81]

The lack of sex-disaggregated data affects our ability to give women sound medical advice. In 2011 the World Cancer Research Fund complained that only 50% of studies into the impact of diet on cancer that included both men and women disaggregated their data by sex, making it hard to establish dietary guidelines for cancer prevention that are valid for both sexes.[82] Women, for example, should probably eat more protein than men as they age (because of muscle mass loss), but 'the optimal dose per meal to support muscle protein synthesis in older women has not been determined'.[83]

The failure to sex-disaggregate when you've actually gone to the effort of including both sexes is baffling, not to mention, as Londa Schiebinger at Stanford University puts it, 'money wasted [and] research that is lost to future meta-analysis'.[84] And when female representation in trials is so low, the ability to conduct meta-analysis can mean the difference between life and death.

In 2014 a review of the FDA database of a cardiac resynchronisation therapy device (CRT-D – essentially a more complicated kind of pacemaker) trials found that women made up about 20% of participants.[85] The number of women included in each individual study was so low that separating out the data for men and women didn't reveal anything statistically significant. But when the review authors combined all the trial results and sex-disaggregated *that* data, they found something alarming.

A CRT-D is used to correct a delay in your heart's electrical signals. They are implanted for established heart failure and the D stands for defibrillator. This defibrillator (a larger version of which most of us will have seen in one hospital drama or other) performs something like a hard reset on the heart, shocking it out of its irregular rhythm so that it can restart in its correct rhythm. A doctor I spoke to described CRT-Ds as 'symptom control'. They aren't a cure, but they prevent many early deaths, and if your heart takes 150 milliseconds or longer to complete a full electrical wave, you should have one implanted. If your heart completes a full circuit in under that time, you wouldn't benefit from one.

Unless, the meta-analysis found, you happened to be female. While the 150 milliseconds threshold worked for men, it was twenty milliseconds too high for women. This may not sound like much, but the meta-analysis found that women with an electrical wave of between 130–49 milliseconds had a 76% reduction in heart failure or death and a 76% reduction in death alone from having the advanced pacemaker implanted. But these women would not be given the device under the guidelines. And so because the trials treated male bodies as the default, and women as a side-show, they had condemned hundreds of women to avoidable heart failure and death.

The CRT-D is far from the only piece of medical tech that doesn't work for women – which is unsurprising given a 2014 analysis which found that only 14% of post-approval medical-device studies included sex as a key outcome measure and only 4% included a subgroup analysis for female participants.[86] A 2010 paper found that 'the female gender is associated with an increased risk of acute complications during primary pacemaker implantation, being independent from age or type of device implanted'.[87] In 2013, a supposedly revolutionary artificial heart was developed that was too big for women.[88] Its designers are working on a smaller version, which is great, but it's striking that, like other artificial hearts,[89] the female version comes years after the default male one.

Even something as basic as advice on how to exercise to keep disease at bay is based on male-biased research. If you run a general search for whether resistance training is good for reducing heart disease, you'll come across a series of papers warning against resistance training if you have high blood pressure.[90] This is in large part because of the concerns that it doesn't have as beneficial an effect on lowering blood pressure as aerobic exercise, and also because it causes an increase in artery stiffness.

Which is all true. In men. Who, as ever, form the majority of research participants. The research that has been done on women suggests that this advice is not gender-neutral. A 2008 paper, for example, found that not only does resistance training lower blood pressure to a greater extent in women, women don't suffer from the same increase in artery stiffness.[91] And this matters, because as women get older, their blood pressure gets higher compared to men of the same age, and elevated blood pressure is more directly linked to cardiovascular mortality in women than in men. In fact, the risk of death from coronary artery disease for women is twice that for men for every 20 mm Hg increase in blood pressure above normal levels. It also matters because commonly used antihypertensive drugs have been shown to be less beneficial in lowering blood pressure in women than in men.[92]

So to sum up: for women, the blood-pressure drugs (developed using male subjects) don't work as effectively, but resistance training just might do the trick. Except we haven't known that because all the studies have been done on men. And this is before we account for the benefits to women in doing resistance training to counteract osteopenia and osteoporosis, both of which they are at high risk for post-menopause.

Other male-biased advice includes the recommendation for diabetics to do high-intensity interval training; it doesn't really help female diabetics[93] (we don't really know why, but this is possibly because women burn fat more than carbs during exercise[94]). We know very little about how women respond to concussions,[95] 'even though women suffer from concussions at higher rates than men and take longer to recover in comparable sports'.[96] Isometric exercises fatigue women less (which is relevant for post-injury rehabilitation) because men and women have different ratios of types of muscle fibre, but we have 'a limited understanding of the differences' because there are 'an inadequate number of published studies'.[97]

When even something as simple as ice-pack application is sex-sensitive, it's clear that women should be included in sports-medicine research at the same rates as men.[98] But they aren't.[99] And researchers continue to research men and act as if their findings apply to women. In 2017, a Loughborough University study[100] was hailed around the UK news media as proving that a hot bath has anti-inflammatory and blood-sugar response benefits similar to exercise.[101] Published in the journal *Temperature* with the sub heading 'A possible treatment for metabolic diseases?' the study included no women at all.

We know that men and women have different metabolic systems. We know that diabetes, one of the diseases particularly singled out as being relevant to this discovery, also affects men and women differently,[102] and that it is a greater risk factor for cardiovascular disease in women than in men.[103] But despite all this, the paper's authors consistently failed to acknowledge any relevance of sex differences to their research. They cited animal studies that had similarly been conducted in all male populations, and perhaps most shockingly of all, in a section specifically looking at 'limitations with the present investigation' they completely failed to mention the fact that the study was all-male as a potential drawback, only referring to their 'relatively small sample size'.

There have been some attempts to force researchers to properly represent females in medical research. Since 1993, when the US passed the National Institute of Health Revitalization Act, it has been illegal not to include women in federally funded clinical trials. Australia's main funding body made similar rules for the research it funds,[104] as has the EU, which in fact went even further, also requiring both sexes to be studied in pre-clinical animal studies. This requirement did not come into effect in the US until January 2016,[105] which is also when the NIH introduced the requirement that the data in trials it funded be disaggregated and analysed by sex (unless there is a compelling reason not to).[106]

Other positive developments include the German Society of Epidemiology which has for more than a decade required researchers to justify including only one sex in any study where the results could potentially affect both sexes;[107] and the introduction of the same by the Canadian Institutes of Health in 2012, as well as mandatory questions about the consideration of sex and gender in the study design. Some academic journals also now insist that papers submitted for publication should provide information about the gender of participants in clinical trials, for example.[108]

Trailing behind everyone is the UK, whose main funders 'make no substantive reference to, or requirements regarding, the consideration of gender in research design and analysis',[109] and despite the at-risk population of women suffering more morbidity and mortality,[110] UK research funding for coronary artery disease in men is far greater than for women. Indeed, such is the dearth of gender-based clinical research from within the UK, that Anita Holdcroft, emeritus professor at Imperial College London, has written that for cardiovascular treatment, 'it is pertinent to use studies from North America and Europe where these issues have been investigated'.[111]

Still, while the situation in the UK is dire, significant problems remain elsewhere. For a start, the evidence we've just seen on the representation of women in trials suggests that these policies are not being rigorously enforced. And, indeed, this is what analyses of the NIH have found. Four years after the NIH announced their first policy calling for the inclusion of women in medical trials, a report was released by the GAO which criticised the NIH for having 'no readily accessible source of data on the demographics of NIH study populations', making it impossible to determine if the NIH was enforcing its own recommendations.[112] By 2015 the GAO was still reporting that the NIH 'does a poor job of enforcing rules requiring that clinical trials include both sexes'.[113]

There also remain plenty of loopholes for US drug manufacturers who don't want the cost and complication of including unharmonious females with their messy hormones in their neat clinical trials, because the rules only apply to NIH-funded trials; independent drug manufacturers can do whatever they want. And the evidence suggests that many of them do: a 2016 paper found that 'a quarter of the drug manufacturers in an industry survey did not deliberately recruit representative numbers of women as participants in drug trials.'[114] When it comes to generic drugs, the FDA only specifies 'guidelines' rather than rules and, as we've seen, these guidelines are being roundly ignored. And the NIH policy on including female subjects in clinical trials doesn't apply to cell studies.

Then of course there's the issue of legacy drugs. Two million women per year take Valium for conditions ranging from anxiety to epilepsy, and it was aggressively marketed towards women for decades.[115] And yet, a 2003 paper points out,[116] this 'mother's little helper' was never tested in randomised clinical trials with female subjects. A 1992 survey by the US General Accounting Office (the Congressional watchdog) found that less than half of publicly available prescription drugs had been analysed for sex differences.[117] A 2015 Dutch paper baldly states that 'The specific effect on women of a huge number of existing medications is simply unknown.'[118]

There is clearly a long way to go, and we must begin to address these gaps as a matter of urgency, because while they remain open, women (who ingest approximately 80% of pharmaceuticals in the US[119]) are dying. Some drugs used to break up blood clots immediately after a heart attack can cause 'significant bleeding problems in women.'[120] Other drugs that are commonly prescribed to treat high blood pressure have been found to lower men's mortality from heart attack – but to increase cardiac-related deaths among women.'[121] Statins, which are regularly prescribed around the world as a preventative measure for heart disease have mainly been tested in men and recent research from Australia suggests that women taking statins at higher dosages may face an increased diabetes risk[122] – which in turn is a higher risk factor for cardiovascular disease in women than in men.[123] In 2000 the FDA forced drug manufacturers to remove phenylpropanolamine, a component of many over-the-counter medications, from all products because of a reported increased risk of bleeding into the brain or into tissue around the brain in women, but not in men.[124] Drug-induced acute liver failure has also been reported more often in women,[125] and certain HIV medications are six to eight times more likely to cause an adverse drug reaction (ADR) in women.[126]

In 2014, the FDA released a database of ADR reports between 2004–13 which showed that women are far more likely than men to experience an ADR: more than 2 million were recorded for women compared to less than 1.3 million for men.[127] Although around the same numbers of men and women die from an ADR, death is ninth on the list of most common ADRs for women, compared to first on the list for men. The second-most common ADR for women (after nausea) is that the drug simply doesn't work at all, and data on the number of deaths that occur as a result of the drug failing to work is not available. We do know, however, that women are more likely to be hospitalised following an ADR,[128] and more likely to experience more than one .[129] A 2001 US study found that 80% of drugs that had been recently removed from the market caused more ADRs in women,[130] while a 2017 analysis points to the 'large number' of medications and medical devices removed from the market by the FDA that posed greater health risks to women.[131]

None of this should surprise us, because despite obvious sex differences, the vast majority of drugs, including anaesthetics and chemotherapeutics,[132] continue with gender-neutral dosages,[133] which puts women at risk of overdose.[134] At a most basic level, women tend to have a higher body-fat percentage than men, which, along with the fact that blood flow to fat tissue is greater in women (for men it's greater to skeletal muscle) can affect how they metabolise certain drugs.[135] Acetaminophen (an ingredient in many pain relievers), for example, is eliminated by the female body at approximately 60% of the rate documented in men.[136] Sex differences in drug metabolism is in part because women's lower lean body mass results in a lower base metabolic rate,[137] but it can also be affected by, among other things: sex differences in kidney enzymes;[138] in bile acid composition (women have less);[139] and intestinal enzyme activity.[140] Male gut transit times are also around half the length of women's, meaning women may need to wait for longer after eating before taking medications that must be absorbed on an empty stomach.[141] Kidney filtering is also faster in men, meaning some renally excreted medications (for example digoxin – a heart medication) 'may require a dosage adjustment'.[142]

For millennia, medicine has functioned on the assumption that male bodies can represent humanity as a whole. As a result, we have a huge historical data gap when it comes to female bodies, and this is a data gap that is continuing to grow as researchers carry on ignoring the pressing ethical need to include female cells, animals and humans, in their research. That this is still going on in the twenty-first century is a scandal. It should be the subject of newspaper headlines worldwide. Women are dying, and the medical world is complicit. It needs to wake up.

CHAPTER 11

Yentl Syndrome

In the 1983 film *Yentl*, Barbra Streisand plays a young Jewish woman in Poland who pretends to be a man in order to receive an education. The film's premise has made its way into medical lore as 'Yentl syndrome', which describes the phenomenon whereby women are misdiagnosed and poorly treated unless their symptoms or diseases conform to that of men. Sometimes, Yentl syndrome can prove fatal.

If I were to ask you to picture someone in the throes of a heart attack, you most likely would think of a man in his late middle age, possibly overweight, clutching at his heart in agony. That's certainly what a Google image search offers up. You're unlikely to think of a woman: heart disease is a male thing. But this stereotype is misleading. A recent analysis of data from 22 million people from North America, Europe, Asia and Australasia found that women from lower socio-economic backgrounds are 25% more likely to suffer a heart attack than men in the same income bracket.[1]

Since 1989, cardiovascular disease has been the leading cause of death in US women and, following a heart attack, women are more likely to die than men.[2] This disparity in deaths has been the case since 1984, and young women appear to be particularly at risk: in 2016 the *British Medical Journal* reported that young women were almost twice as likely as men to die in hospital.[3] This may be in part because doctors aren't spotting at-risk women: in 2016, the American Heart Association also raised concerns about a number of risk-prediction models 'commonly used' in patients with acute coronary syndrome, because they were developed in patient populations that were at least two-thirds male.[4] The performance of these risk-prediction models in women 'is not well established'.

Common preventative methods may also not work as well in women. Acetylsalicylic acid (aspirin) has been found to be effective in preventing a first heart attack in men, but a 2005 paper found that it had a 'nonsignificant' effect in women aged between forty-five and sixty-five.[5] Prior to this study, the authors noted, there had been 'few similar data in women'. A more recent study from 2011 found that not only was aspirin ineffective for women, it was potentially harmful 'in the majority of patients'.[6] Similarly, a 2015 study found that taking a low dose of aspirin every other day 'is ineffective or harmful in the majority of women in primary prevention' of cancer or heart disease.[7]

Perhaps the greatest contributor to the numbers of women dying following a heart attack, however, is that their heart attacks are simply being missed by their doctors. Research from the UK has found that women are 50% more likely to be misdiagnosed following a heart attack (rising to almost 60% for some types of heart attack[8]). This is partly because women often don't have the 'Hollywood heart attack' as it's known in medical circles (chest and left-arm pains).[9] Women (particularly young women) may in fact present without any chest pain at all, but rather with stomach pain, breathlessness, nausea and fatigue.[10] These symptoms are often referred to as 'atypical', a designation to which the *British Medical*

Journal took exception in a 2016 article, saying that the term 'may lead to the under-appreciation of risk associated with this presentation'.[11] And under appreciation of the risk may in turn explain why a 2005 US study found that 'only one in five physicians across multiple specialties was aware that more women than men die from cardiovascular disease each year, and most of these physicians did not rate themselves as effective in treating sex-tailored cardiovascular disease'.[12]

Atypical or not, for certain types of heart attacks, women (and again especially young women) who present without chest pain are at particular risk of death[13] – which makes it extremely concerning that current NHS England guidelines specify 'acute cardiac sounding chest pain' as part of the criteria for a patient being referred for primary percutaneous coronary interventions (PPCI) at one of the country's specialist twenty-four-hour heart-attack centres.[14] PPCI is an emergency treatment that restores blood flow during a heart attack, and which according to one doctor I spoke to has 'massively improved survival and outcome'. But this treatment is only carried out at the twenty-four-hour heart-attack centres and, perhaps as a result, 75% of those who receive this treatment are men.[15]

The tests doctors use to determine what's wrong with a patient are also likely contributing to women's higher death rates following a heart attack. Standard tests like the electrocardiogram or the physical stress test have been found to be less conclusive in women.[16] A 2016 *BMJ* paper refers to recent work from Edinburgh which showed that the 'normal' diagnostic threshold for troponin (a protein released into the blood during heart damage) may be too high for women.[17] And it's not just about 'standard' levels for biomarkers being incorrect in women, we also need to establish new female-specific biomarkers.[18] A biomarker is a biological characteristic (like troponin) whose presence can act as a diagnostic criteria for a specific disease, and a 2014 literature review of sex difference studies suggests that this may be a fruitful area to research.[19] Unfortunately, it concludes that the work done so far is too limited to be able to say whether or not female-specific biomarkers will be found.

Because women's heart attacks may not only present differently, but may in fact be mechanically different, the technology we've developed to search for problems may not be suitable for female hearts.[20] For example, a heart attack is traditionally diagnosed with an angiogram, which will show where there are obstructed arteries.[21] But women often don't have obstructed arteries, meaning that the scan won't show up any abnormalities,[22] and women who turn up at hospital with angina (chest pain) may simply be discharged with a diagnosis of 'non-specific chest pain' and told they have no significant disease.[23] Except they do: women with 'normal' angiograms have gone on to suffer a heart attack or stroke shortly after being discharged from hospital.[24]

Assuming a woman gets lucky and has her heart disease diagnosed, she must then navigate the obstacle course of male-biased treatment: sex differences have not generally been integrated either into 'received medical wisdom' or even clinical guidelines.[25] For example, say a man and a woman are both diagnosed with a swollen aorta (the aorta is the main blood vessel that runs from the heart down through the chest and stomach). They are both suffering from an equal level of swelling – but their risk is not the same: the woman has a higher risk of rupture, which carries with it a 65% chance of death.[26] And yet, in Dutch clinical guidelines, the thresholds for surgery don't differ for each sex.[27]

Diagnostic tests developed around male bodies are also a problem in other medical disciplines, even those where women are more at risk. Women have a higher risk than men of developing right-sided colon cancer, which often develops more aggressively,[28] but the faecal blood test commonly used to detect colon cancer is less sensitive in women than in men.[29] Meanwhile, because women have on average a longer and narrower colon than men, colonoscopies in women may be incomplete.[30] Then there's what the WHO calls the 'frequent mistake' of underestimating the importance of symptoms that can only occur in one sex, such as vaginal bleeding in dengue fever.[31] When symptoms are listed in order of frequency for all patients rather than separated by sex, female-specific symptoms can be presented as less significant than they are in reality.

The impact of such data gaps can snowball. When it comes to tuberculosis (TB), for example, a failure to account for how female social roles could make the disease more dangerous for women combines with a failure to collect sex-disaggregated data, leading to potentially deadly consequences.[32] Men are more likely to have latent TB, but women are more likely to develop the active disease.[33] Studies also suggest that women in developing countries who cook in poorly ventilated rooms with biomass fuels (as we've seen, this means millions of women) have impaired immune systems which leave them less able to fight off the bacteria.[34] The result is that TB kills more women globally than any other single infectious disease. More women die annually of TB than of all causes of maternal mortality combined.[35] But TB is nevertheless often considered to be a 'male disease', and as a result women are less likely to be screened for it.

Even when women are screened, they are less likely to be diagnosed.[36] Women may have a different immune response to TB resulting in different symptoms,[37] and one study on why women are misdiagnosed found that TB lung lesions might not appear as severe in women.[38] There is also evidence of sex differences in the sensitivity of commonly used screening tests.[39] The standard way to test for TB in resource-limited settings is to get patients to cough up sputum and examine it under the microscope.[40] But women with TB are less likely to have a sputum-producing cough, and even if they do have one their sputum is less likely to test positive for the disease.[41] The sputum test is also problematic for social reasons: a study in Pakistan reported that women felt uncomfortable coughing up the mucus needed for the examination, and health workers weren't explaining why they needed to. So they didn't.[42]

Medical practice that doesn't account for female socialisation is a widespread issue in preventative efforts as well. The traditional advice of using condoms to avoid HIV infection is simply not practicable for many women who lack the social power to insist on their use. This also goes for Ebola, which can remain present in semen for up to six months. And although a gel has been developed to address this problem,[43] it fails to account for the practice of 'dry sex' in certain parts of sub-Saharan Africa.[44] A gel which also acts as a lubricant will not be acceptable in areas where women de-lubricate their vaginas with herbs in order to indicate that they are chaste.

Failing to account for female socialisation can also lead to women living for decades with undiagnosed behavioural disorders. For years we have thought that autism is four times more common in boys than in girls, and that when girls have it, they are more seriously affected.[45] But new research suggests that in fact female socialisation may help girls mask their symptoms better than boys and that there are far more girls living with autism than we previously realised.[46] This historical failure is partly a result of the criteria for diagnosing autism having been based on data 'derived almost entirely' from studies of boys,[47] with a 2016 Maltese study concluding that a significant cause of misdiagnosis in girls was 'a general male-bias in diagnostic methods and clinical expectations'.[48] There is also emerging evidence that some girls with anorexia may in fact be suffering from autism, but because it's not a typical male symptom it's been missed.[49] Sarah Wild, head of Limpsfield Grange, the UK's only state-funded residential school for girls with special needs, told the *Guardian* that 'the diagnostic checklists and tests have been developed for boys and men, while girls and women present completely differently'.[50] Meanwhile, a recently published draft of new NHS guidance on autism made no mention of women's differing needs.[51]

There are similar diagnostic problems when it comes to attention deficit hyperactivity disorder and Asperger's. A 2012 survey by the UK's National Autistic Society found that just 8% of girls with Asperger's syndrome were diagnosed before the age of six, compared with 25% of boys; by the age of eleven the figures were 21% and 52%, respectively.[52] Up to three-quarters of girls with ADHD are estimated to be undiagnosed – a gap which Dr Ellen Littman, the author of *Understanding Girls with ADHD*, puts down to the early clinical studies of ADHD having been done on 'really hyperactive young white boys'. Girls tend to present less as hyperactive and more as disorganised, scattered and introverted.[53]

More broadly, researchers suggest that because women are socialised to 'take turns in conversation, to downplay their own status, and to demonstrate behaviors that communicate more accessibility and friendliness', the traditional medical interview model may be unsuccessful in getting the information from women that is needed to diagnose them effectively.[54] But sometimes – often – women are providing the information. It's just that they aren't being believed.

American news website *ThinkProgress* reported the story of Kathy, whose heavy periods left her feeling so faint she couldn't stand.[55] But when it came to getting a diagnosis, Kathy faced the same problem encountered by Michelle in the previous chapter. Four different medical professionals thought it was in her head, that 'she was simply struggling with anxiety and perhaps even had a serious mental health disorder'. Her primary-care doctor went so far as to tell her more than once, 'All your symptoms are in your imagination.'

But they weren't in her imagination. In fact, Kathy turned out to have 'potentially life-threatening uterine fibroids that required surgical intervention', something that was only discovered after she demanded an ultrasound. She wasn't anxious (although after nine months of being told she was crazy who could blame her if she was), she was anaemic.

Rachael was also told she was imagining it. She had been trying to manage her severe pain and heavy periods with the pill for ten years by the time she collapsed at a gig. The hospital sent her home with painkillers and a diagnosis of stress. The next time she collapsed the hospital put her in the gastroenterology ward. 'Six nights I was there, on a drip. There was a woman dying of bowel cancer in the bed opposite me. It was horrible.' The doctors suspected kidney stones, so they ran multiple tests around her urinary system. They all came back negative. So did her blood tests. And the more tests that came back negative, the more Rachael sensed a shift in how she was being treated. 'I started feeling they weren't believing me. That they thought it was all in my head.' Eventually a consultant shook his head as Rachael told him how much she hurt and told her, 'We have to send you home. There's nothing wrong with you.'

But there was something wrong with her. Rachael was eventually diagnosed with endometriosis, a disease where womb tissue grows elsewhere in the body, causing extreme pain and sometimes infertility. It takes an average of eight years to diagnose in the UK,[56] an average of ten years to diagnose in the US,[57] and there is currently no cure. And although the disease is thought to affect one in ten women (176 million worldwide[58]) it took until 2017 for England's National Institute for Health and Care Excellence to release its first ever guidance to doctors for dealing with it. The main recommendation? 'Listen to women.'[59]

This may be easier said than done, because failing to listen to female expressions of pain runs deep, and it starts early. A 2016 study from the University of Sussex played a series of cries to parents (twenty-five fathers and twenty-seven mothers) of three-month-old babies. They found that although babies' cries aren't differentiated by sex (sex-based pitch differences don't occur until puberty) lower cries were perceived as male and higher cries perceived as female. They also found that when male parents were told that a lower-pitched cry belonged to a boy, they rated the baby as in more discomfort than when the cry was labelled female.

Instead of believing women when they say they're in pain, we tend to label them as mad. And who can blame us? Bitches be crazy, as Plato famously said. Women are hysterical (*hystera* is the Greek word for womb), crazy (if I had a pound for every time a man questioned my sanity in response to my saying anything vaguely feminist on Twitter I would be able to give up work for life), irrational and over emotional. The trope of the 'crazy ex-girlfriend' is so common it's been satirised by Taylor Swift in her hit song 'Blank Space' and by Rachel Bloom in a whole Netflix series about a *Crazy Ex-Girlfriend*. Women are a 'mystery', explained renowned physicist Stephen Hawking,[60] while Freud, who got rich and famous off his diagnoses of female hysteria, explained in a 1933 lecture that 'Throughout history, people have knocked their heads against the riddle of femininity.'[61]

The intransigence of this feminine riddle has not gone unpunished. Women who had often done little more than manifest behaviours that were out of feminine bounds (such as having a libido) were incarcerated for years in asylums. They were given hysterectomies and clitoridectomies. Women were locked up for having even mild post-natal depression: the grandmother of a friend of mine spent her life in an asylum after throwing a scourer at her mother-in-law. At least one US psychiatric textbook, still widely in use during the 1970s, recommended lobotomies for women in abusive relationships.[62]

Of course, we've moved on from such inhumane treatment of women. We no longer lock women up and cut out parts of their brains. Instead, we give women drugs: women are two and a half times more likely to be on antidepressants than men.[63] This is not to condemn antidepressants: they can be life-changing for people with mental health problems. However, it's still worth asking *why* women are so much more likely to be on them, because it's not simply that women are more likely to seek help. A 2017 Swedish study in fact found that it was men who were more likely to report depression.[64] So why are more women being treated with antidepressants? Are women simply more 'feeble-minded'? Does living in a world in which we don't quite fit affect our mental health? Or are antidepressants the new (and obviously preferable) lobotomy for women dealing with trauma?

Freud once believed that hysteria might be linked to historic sexual abuse. He later retracted this theory as it would have implicated too many men to be, in his opinion, credible. But recent research suggests that abuse might be linked to certain types of pain women experience[65] – and in the wake of the #MeToo global scandal maybe it's not so incredible after all.

The full answer to these questions is beyond the scope of this book. But one possible explanation for at least part of the disparity is that women are being prescribed antidepressants when they are not in fact depressed. Women's physical pain is far more likely to be dismissed as 'emotional' or 'psychosomatic'. The Swedish study which found that men are more likely to report depression also found that women who have not reported depression are twice as likely as men to be prescribed antidepressants. This chimes with studies from the 1980s and 90s which found that while men who reported pain tended to receive pain medication, women were more likely to receive sedatives or antidepressants.[66] A 2014 study which required healthcare providers to make treatment recommendations for hypothetical patients with lower back pain similarly found that female patients were significantly more likely to be prescribed antidepressants than men.[67]

It seems that Yentl syndrome may be at play again here: it is striking that so many of the stories women tell of undiagnosed and untreated pain turn out to have physical causes that are either exclusively female diseases, or are more common in women than in men. Women are almost twice as likely to have irritable bowel syndrome as men[68] and three times more likely to experience migraines[69] (a condition about which we know very little despite it being chronic, often deeply debilitating and affecting 37 million Americans[70] and one in eight people in the UK[71]). In fact, many clinical pain conditions are substantially more prevalent in women than men,[72] and several studies over the past decades have shown that women are more sensitive to pain than men (which sheds a particularly cruel light on the finding that women are less likely to receive painkillers).

There is also mounting evidence that men and women may experience pain differently. A woman's pain sensitivity increases and decreases throughout her menstrual cycle, 'with skin, subcutaneous tissue, and muscles being affected differently by female hormonal fluctuations'.[73] An animal study which found that males and females use different types of immune cells to convey pain signals may provide the beginnings of an answer as to why[74] – although only the beginnings: sex differences in pain remain an under-researched area and even what we do know is not widely dispersed. Dr Beverly Collett, who until she retired in 2015 was a consultant at Leicester's pain management service and chair of the Chronic Pain Policy Coalition, told the *Independent* that the average GP 'has no idea that drugs such as paracetamol and morphine work differently in women'.[75]

Even if they are treated for their pain, women routinely have to wait longer than men to receive that treatment. A US analysis of

92,000 emergency-room visits between 1997 and 2004 found women had longer waiting times than men,[76] and a study of adults who presented to a US urban emergency department between April 2004 and January 2005 found that while men and women presented with similar levels of pain, women were less likely to receive analgesia and women who did receive analgesia waited longer to receive it.[77] A US Institute of Medicine publication on chronic pain released in 2011 suggested that not much has changed, reporting that women in pain face 'delays in correct diagnosis, improper and unproven treatments', and 'neglect, dismissal and discrimination' from the healthcare system.[78] In Sweden a woman suffering from a heart attack will wait one hour longer than a man from the onset of pain to arrival at a hospital, will get lower priority when waiting for an ambulance, and will wait twenty minutes longer to be seen at the hospital.[79]

The reality that female bodies are simply not afforded the same level of medical attention as male bodies is often brushed aside with the riposte that, on average, women enjoy more years of life than men. But while it is true that female life expectancy remains a few years longer than male life expectancy (although that gap is narrowing as women's lives have become less prescriptive and occupational safety in male-dominated jobs has become more stringent), there is evidence to suggest that the female mortality advantage isn't exactly secure.

A 2013 paper that examined trends in US mortality rates from 1992–2006 in 3,140 counties reported that even as mortality decreased in most counties, female mortality increased in 42.8% of them.[80] And while men's years of good health have increased in line with their longevity, both women's longevity and active years have increased at a much lower rate: thirty years of US health data showed that, while women live on average five years longer than men (in Europe it is 3.5 years[81]), those years are spent in ill health and disability.[82]

The result is that US women no longer have more active years than men,[83] despite their longer lives, and while women account for 57% of US citizens aged over sixty-five, they make up 68% of those who need daily assistance.[84] In 1982 both men and women who lived to eighty-five could expect two and a half further years of active healthy life. For women, that figure hasn't changed, but an eighty-five-year-old man alive now can expect to be active and healthy until he's eighty-nine. The trend of increasing longevity and good health amongst men can also be found in Belgium[85] and Japan.[86] A WHO paper into women's health in the EU reported that in 2013, 'even in countries with some of the highest overall life expectancy in the Region, women spent almost 12 years of their life in ill health'.[87] And, yes, it would be nice to have some sex-disaggregated data on why this is happening.

A particularly troubling side effect of Yentl syndrome is that when it comes to medical issues that mainly or only affect women, you can forget about including women in trials because here the research is often lacking altogether.

Premenstrual syndrome (PMS) is a collection of symptoms that can include among other things: mood swings, anxiety, breast tenderness, bloating, acne, headaches, stomach pain and sleep problems. PMS affects 90% of women, but is chronically under-studied: one research round-up found five times as many studies on erectile dysfunction than on PMS.[88] And yet while a range of medication exists to treat erectile dysfunction[89] there is very little available for women, to the extent that over 40% of women who have PMS don't respond to treatments currently available. Sufferers are still sometimes treated with hysterectomies; in extreme cases, women have tried to kill themselves.[90] But researchers are still being turned down for research grants on the basis that 'PMS does not actually exist'.[91]

Period pain – dysmenorrhea – similarly affects up to 90% of women,[92] and according to the American Academy of Family

Physicians it affects the daily life of around one in five women.[93] The level of pain women experience on a monthly basis has been described as 'almost as bad as a heart attack'.[94] But despite how common it is and how bad the pain can be, there is precious little that doctors can or will do for you. A rare 2007 grant application for research into primary dysmenorrhea described its causes as 'poorly understood' and treatment options as 'limited'.[95] The prescription medications which are available have serious possible side effects and are by no means universally effective.

When I went to my (male) doctor about period pain that wakes me up at night and leaves me in a moaning foetal position in the daytime, he more or less laughed me out of the room. I haven't bothered trying again. So imagine my joy when I read about a 2013 study that seemed to have found a cure. The 'primary outcome' of a double-blind, randomised, controlled trial of sildenafil citrate, was, ladies, you may want to sit down for this: 'total pain relief over 4 consecutive hours', with 'no observed adverse effects'.[96] Imagine.

Created in 1989, sildenafil citrate is the medical name for Viagra. In the early 1990s, the drug was being tested as a heart-disease medication.[97] It turned out not to be great at that, but one thing participants *did* report was an increase in erections (yes, all the trial participants were men). Total erectile dysfunction affects between 5–15% of men depending on age,[98] with about 40% of men experiencing it to some degree – and so naturally the researchers were keen to explore this alternative use for their drug. By 1996, sildenafil citrate had been patented in the US and by March 1998 it was approved by the FDA.[99] A happy ending for men, then.

But what if the trial had included women? The outcome of the 2013 study is suggestive. The trial had to be stopped because the funding ran out, meaning the researchers did not meet their sample size and therefore could not confirm the primary hypothesis. They called for 'larger studies of longer duration, likely multi-center' to confirm their findings.

These studies have not happened. Dr Richard Legro, who led the study, told me he applied twice to the NIH for funding 'to do a longer and larger study and also to compare sildenafil to the standard of care, a non-steroidal anti-inflammatory agent'. He was rejected both times. In each case, the grant 'was deemed to be in the lower half of grants submitted'. It wasn't even reviewed. Legro tells me that the comments he received 'indicated that the reviewers did not see dysmenorrhea as a priority public health issue'. They also didn't 'fully understand clinical trial design of dysmenorrhea trials'. When I ask him if he thinks he will ever get funding, he says, 'No. Men don't care or understand dysmenorrhea. Give me an all-female review panel!'

The failure of pharmaceutical companies to step in here and capitalise on what is surely a gold-plated commercial opportunity may seem baffling, but it's quite possibly just another data-gap problem. In an email, Legro told me that, for cost reasons, the pharma industry 'doesn't usually fund investigator-initiated projects', particularly of drugs that are available generically. And this may be where the data gap comes in: there simply isn't much research done on dysmenorrhea,[100] which makes it difficult for pharma companies to know exactly how much money could be made on such a drug – and therefore makes it harder for them to decide to fund trials. Especially if the people making the decisions happen not to be women. Legro also suggested that pharma companies may not want to risk doing tests in women in case of negative findings that would endanger the use of sildenafil in men. In short, it seems that pharma companies may in fact *not* see this as a gold-plated commercial opportunity. And so women carry on being incapacitated by pain on a monthly basis.

Male-dominated funding panels may also explain why we have so few drugs available for uterine failure. Every day 830 women around the world die due to complications during pregnancy and childbirth[101] (in some African countries more women die annually from childbirth than at the height of the Ebola epidemic[102]). Over half of these deaths are explained as being a result of problems with contractions, often because the contractions are too weak for the woman to give birth. The only medical treatment available for women whose contractions aren't strong enough is the hormone oxytocin, which works about 50% of the time. Those women go on to give birth vaginally. Women who don't respond to oxytocin are given an emergency caesarean. In the UK weak contractions are the reason given for a majority of the 100,000 emergency caesareans carried out each year.

We currently have no way of knowing which women will respond to oxytocin, which clearly isn't ideal: all women, including those for whom it will result in a pointless and harrowing delay, have to go through the process. This happened to a friend of mine in 2017. After being in hospital in excruciating pain for two days (on her own for much of it as her partner had been sent home), she was only 4 cm dilated. Eventually she was taken off for a C-section, and she and the baby were fine. But the experience left her traumatised. She had flashbacks for the first few weeks after she gave birth. When she talks about the internal exams and procedures, she describes it as a violent assault. It was, she says, brutal. But what if it didn't have to be this way? What if they'd known from the beginning that she was going to need a caesarean?

In 2016 Susan Wray, a professor of cellular and molecular physiology at the University of Liverpool, gave a lecture to the Physiological Society.[103] Wray is also the director of the Centre of Better Births in Liverpool Women's Hospital and she explained that recent research revealed that women with contractions that were too weak to give birth had more acid in their myometrial blood (the blood in the part of the uterus that causes contractions). The higher the levels of acid were, the higher the likelihood a woman would end up needing a caesarean, because oxytocin isn't, it turns out, that effective on women with an acidic blood pH.

But Wray didn't simply want to be able to predict the need for a caesarean. She wanted to be able to avoid it. Together with her fellow researcher Eva Wiberg-Itzel, Wray conducted a randomised control trial on women with weak contractions. Half of them were given the usual oxytocin; half were given bicarbonate of soda, and then given the usual oxytocin an hour later. The change was dramatic: 67% of women given just oxytocin went on to give birth vaginally, but this rose to 84% if they were given bicarbonate of soda an hour before. As Wray pointed out, the bicarb dose wasn't tailored to body weight, it wasn't tailored to the amount of acid in the blood, and the women weren't given repeated doses. So the efficacy could turn out to be even higher.

This finding could not only be transformative for the tens of thousands of women a year who have what could turn out to be unnecessary surgery (not to mention saving the NHS a substantial amount of money). It could save women's lives in countries where caesarean sections are risky or not readily available – not that you have to live in a low-income country for a C-section to be risky: you could just be a black woman living in the United States.[104]

The US has the highest maternal mortality rate in the developed world, but the problem is particularly acute for African Americans. The World Health Organization has estimated that the death rate of black expectant and new mothers in the US matches that of women in much lower-income countries like Mexico and Uzbekistan. Black women in America have worse health outcomes overall than white women, but when it comes to pregnancy and childcare the comparisons score off the charts: African American women are

243% more likely than white women to die from pregnancy and childbirth-related issues. And it's not just because African Americans tend to be poorer: a 2016 analysis of births in New York City found that 'black college-educated mothers who gave birth in local hospitals were more likely to suffer severe complications of pregnancy or childbirth than white women who never graduated from high school'. Even global tennis superstar Serena Williams is not immune: in February 2018 she revealed that she had almost died following an emergency C-section.[105] African American women also have higher rates of caesarean section and a 2015 study from Connecticut found that – even when controlling for socio-economic status – black women were more than twice as likely to have to return to hospital in the month following surgery.[106] So Wray's research could be transformative here.

But it looks like we aren't going to see the fruits of her labour any time soon. When Wray discovered that the British Medical Research Council was offering funding for research that would benefit low- and middle-income countries, she decided to apply. And yet, despite all the data about how dangerous weak contractions can be, she was turned down. The research was 'not a high enough priority'. So currently we have only one treatment for women with weak contractions, and it doesn't work half the time. Compare this, Wray says, to the around fifty drugs available for heart failure.

The evidence that women are being let down by the medical establishment is overwhelming. The bodies, symptoms and diseases that affect half the world's population are being dismissed, disbelieved and ignored. And it's all a result of the data gap combined with the still prevalent belief, in the face of all the evidence that we *do* have, that men are the default humans. They are not. They are, to state the obvious, just men. And data collected on them does not, cannot, and should not, apply to women. We need a revolution in the research and the practice of medicine, and we need it yesterday. We need to train doctors to listen to women, and to recognise that their inability to diagnose a woman may not be because she is lying or being hysterical: the problem may be the gender data gaps in their knowledge. It's time to stop dismissing women, and start saving them.

PART V

Public Life

CHAPTER 12

A Costless Resource to Exploit

'How much will it cost?' This is the first question that anyone proposing a policy initiative must answer, swiftly followed by 'Can we afford it?' The answer to the first question will be fairly straightforward, but the answer to the second is a little trickier. It will depend on the current state of a country's economy, and that figure is more subjective than many of us think.

The standard measure of a country's economy is gross domestic product (GDP) and if economics has a religion, then this is its god. It is compiled from data collected in a range of surveys and represents the total value of goods (how many shoes were manufactured) and services (how many meals were served at restaurants) a country produces. It also includes how much we all got paid and how much we (including governments and businesses) have all spent. It all sounds very scientific, but the truth is that GDP has a woman problem.

The formulation of a country's official GDP figure is an inherently subjective process, explains Diane Coyle, professor of economics at Manchester University. 'A lot of people think that [GDP] is a real thing. But actually, it's a confection, with lots of judgments that have gone into its definition. And a lot of uncertainty.' Measuring GDP is, she says, 'not like measuring how high the mountain is'. When you see headlines proclaiming that 'GDP went up 0.3% this quarter', she cautions, you should remember that that 0.3% 'is dwarfed by the amount of uncertainty in the figures'.

Compounding this uncertainty are glaring gaps in the data used to compile the figures. There are plenty of goods and services that GDP simply doesn't account for. And the decision over which to include is somewhat arbitrary. Until the 1930s we didn't really measure the economy with any seriousness. But that changed in the wake of the Great Depression. In order to address the economic meltdown, governments needed to know more precisely what was going on, and in 1934 a statistician called Simon Kuznets produced the United States' first national accounts.[1] This was the birth of GDP.

Then the Second World War came along, and it was during this period, explains Coyle, that the frame we use now was established. It was designed to suit the needs of the war economy, she tells me. 'The main aim was to understand how much output could be produced and what consumption needed to be sacrificed to make sure there was enough available to support the war effort.' To do this they counted everything produced by government and businesses and so 'what governments do and what businesses do came to be seen as the definition of the economy'. But there was one major aspect of production that was excluded from what came to be the 'international convention about how you think about and measure the economy', and that was the contribution of unpaid household work, like cooking, cleaning and childcare. 'Everyone acknowledges that there is economic value in that work, it's just not part of 'the economy',' says Coyle.

This was not a mere oversight: it was a deliberate decision, following a fairly vigorous debate. 'The omission of unpaid services of housewives from national income computation distorts the picture', wrote economist Paul Studenski in his classic 1958 text *The Income of Nations*. In principle, he concluded, 'unpaid work in the home should be included in GDP'. But principles are man-made, and so 'after a bit of to-ing and fro-ing', and much debate over how you would measure and value unpaid household services 'it was decided', says Coyle, 'that this would be too big a task in terms of collecting the data'.

Like so many of the decisions to exclude women in the interests of simplicity, from architecture to medical research, this conclusion could only be reached in a culture that conceives of men as the default human and women as a niche aberration. To distort a reality you are supposedly trying to measure makes sense only if you don't see women as essential. It makes sense only if you see women as an added extra, a complicating factor. It doesn't make sense if you're talking about half of the human race. It doesn't make sense if you care about accurate data.

And excluding women does warp the figures. Coyle points to the post-war period up to about the mid-1970s. This 'now looks like a kind of golden era of productivity growth', Coyle says, but this was to some extent a chimera. A large aspect of what was actually happening was that women were going out to work, and the things that they used to do in the home – which weren't counted – were now being substituted by market goods and services. 'For example buying pre-prepared food from the supermarket rather than making it from scratch at home. Buying clothes rather than making clothes at home.' Productivity hadn't actually gone up. It had just shifted, from the invisibility of the feminised private sphere, to the sphere that counts: the male-dominated public sphere.

The failure to measure unpaid household services is perhaps the greatest gender data gap of all. Estimates suggest that unpaid care work could account for up to 50% of GDP in high-income countries, and as much as 80% of GDP in low-income countries.[2] If we factor this work into the equation, the UK's GDP in 2016 was around $3.9 trillion[3] (the World Bank's official figure was $2.6 trillion[4]), and India's 2016 GDP was around $3.7 trillion[5] (compared to the World Bank's figure of $2.3 trillion).

The UN estimates that the total value of unpaid childcare services in the US was $3.2 trillion in 2012, or approximately 20% of GDP (valued at $16.2 trillion that year).[6] In 2014 nearly 18 billion hours of unpaid care were provided to family members with Alzheimer's (close to one in nine people aged sixty-five and older in the US are diagnosed with the disease). This work has an estimated value of $218 billion,[7] or, as an *Atlantic* article put it, 'nearly half the net value of Walmart's 2013 sales'.[8]

In 2015, unpaid care and domestic work in Mexico was valued at 21% – 'higher than manufacturing, commerce, real estate, mining, construction and transportation and storage'.[9] And an Australian study found that unpaid childcare should in fact be regarded as Australia's largest industry generating (in 2011 terms) $345 billion, or 'almost three times the financial and insurance services industry, the largest industry in the formal economy'.[10] Financial and insurance services didn't even make second place in this analysis; they were shunted into a lowly third place by 'other unpaid household services'.

You will notice that these are all estimates. They have to be, because no country is currently systematically collecting the data. And it's not because there is no way of doing it. The most common way of measuring the amount of unpaid work women do is with time-use surveys. Individuals are asked to keep a time diary of their movements throughout the day – what they are doing, where, and with whom. It is because of this form of data capture, writes prize-winning economist Nancy Folbre, that we now know that 'in virtually every country, women undertake a disproportionate share of all non-market work, and also tend to work longer hours overall than men do'.

Standard time-use surveys were primarily designed to measure explicit activities such as meal preparation, house-cleaning or feeding a child.[11] As a result, they often fail to capture on-call responsibilities, such as having to keep an eye on a sleeping child or be available for an adult with a serious illness while you get on with something else – another data gap. Time-use surveys that explicitly aim to capture such responsibilities demonstrate that the market value of 'on-call care', even at a very low replacement wage, is significant,[12] but like with travel data this kind of care work is often lost within personal and leisure data.[13] Folbre points to studies of home-based care for HIV/AIDS in Botswana which 'estimated the value of services per caregiver at about $5,000 per year, a number that would substantially increase estimates of total spending on healthcare if it were included'.[14]

The good news is that these surveys have been on the increase in many countries. 'In the first decade of the 21st century, more than 87 such surveys were conducted, more than the total in the entire 20th century', writes Folbre. But reliable time-use information is still lacking for many countries around the world.[15] And measuring women's unpaid work is still seen by many as an optional extra:[16] Australia's scheduled 2013 time-use survey was cancelled, meaning that the most recent Australian data available is from 2006.[17]

Coyle tells me that she 'can't help being a bit suspicious that the original decision not to bother counting work in the home was informed by gender stereotypes in the 1940s and 50s'. Her suspicion seems entirely justified, and not just because the original rationale for excluding women's work was so flimsy. With the rise of digital public goods like Wikipedia and open-source software (which are displacing paid goods like encyclopaedias and expensive proprietary software), unpaid work is starting to be taken seriously as an economic force – one that should be measured and included in official figures. And what's the difference between cooking a meal in the home and producing software in the home? The former has largely been done by women, and the latter is largely done by men.

The upshot of failing to capture all this data is that women's unpaid work tends to be seen as 'a costless resource to exploit', writes economics professor Sue Himmelweit.[18] And so when countries try to rein in their spending it is often women who end up paying the price.

Following the 2008 financial crash, the UK has seen a mass cutting exercise in public services. Between 2011 and 2014 children's centre budgets were cut by £82 million and between 2010 and 2014, 285 children's centres either merged or closed.[19] Between 2010 and 2015 local-authority social-care budgets fell by £5 billion,[20] social security has been frozen below inflation and restricted to a household maximum, and eligibility for a carers' allowance depends on an earnings threshold that has not kept up with increases in the national minimum wage.[21] Lots of lovely money-saving.

The problem is, these cuts are not so much savings as a shifting of costs from the public sector onto women, because the work still needs to be done. By 2017 the Women's Budget Group estimated[22] that approximately one in ten people over the age of fifty in England (1.86 million) had unmet care needs as a result of public spending cuts. These needs have become, on the whole, the responsibility of women.

Cuts have also contributed to a rise in female unemployment: by March 2012, two years into austerity, women's unemployment had risen by 20% to 1.13 million, the highest figure for twenty-five years.[23] Meanwhile, male unemployment stood at almost exactly where it had since the end of the recession in 2009. *Unison* found that by 2014 there had been a 74% increase in women's underemployment.[24]

In 2017 the House of Commons library published an analysis of the cumulative impact of the government's 'fiscal consolidation' between 2010 and 2020. They found that 86% of cuts fell on women.[25] Analysis by the Women's Budget Group (WBG)[26] found that tax and benefit changes since 2010 will have hit women's incomes twice as hard as men's by 2020.[27] To add insult to injury, the latest changes are not only disproportionately penalising poor women (with single mothers and Asian women being the worst affected[28]), they are benefiting already rich men. According to WBG analysis, men in the richest 50% of households actually gained from tax and benefit changes since July 2015.[29]

So why is the UK government enacting policy that is so manifestly unjust? The answer is simple: they aren't looking at the data. Not only are they not quantifying women's unpaid contribution to GDP, the UK government (like most governments worldwide) also aren't gender-analysing their budgets.

By repeatedly refusing (most recently in December 2017) to produce a comprehensive equality impact assessment of its budgets, the UK government has arguably been operating illegally since the public sector equality duty (PSED) came into law. Part of the 2010 Equality Act, the PSED requires that 'a public authority, must, in the exercise of its functions, have due regard to the need to eliminate discrimination [and] advance equality of opportunity'.[30] In an interview with the *Guardian*, WBG's director, Eva Neitzert, couldn't see how the Treasury could fulfil its legal obligations without completing a formal assessment.[31] Were Treasury ministers 'deliberately seeking to hide inconvenient truths about the impact of its policies on women?' she wondered.

If they were, it would be profoundly foolish, because spending cuts on public services are not just inequitable, they are counterproductive. Increasing the amount of unpaid work women have to do lowers their participation rate in the paid labour force. And women's paid labour-force participation rate has a significant impact on GDP.

Between 1970 and 2009, almost 38 million more women joined the US labour force, increasing the female participation rate from 37% to nearly 48%. McKinsey calculates that without this increase, US GDP would be 25% smaller – 'an amount equal to the combined GDP of Illinois, California and New York'.[32] The World Economic Forum (WEF) has also found that increasing female labour participation 'has been an important driver of European economic growth in the last decade'. By contrast, 'Asia and the Pacific reportedly loses US$42 billion to US$47 billion annually as a region because of women's limited access to employment opportunities'.[33]

There are still further gains that could be made. There is a 12% employment gap between men and women across the EU (the figure varies between 1.6% in Latvia and 27.7% in Malta);[34] a 13% gap in the US;[35] and a 27% gap worldwide.[36] The WEF has calculated that closing this gap 'would have massive economic implications for developed economies, boosting US GDP by as much as 9% and eurozone GDP by as much as 13%'.[37] In 2015 McKinsey estimated that global GDP would grow by $12 trillion were women able to engage in the paid labour force at the same rate as men.[38]

But they aren't, because they simply don't have the time. Both the OECD[39] and McKinsey[40] have uncovered a 'strong negative correlation' between time spent in unpaid care work and women's paid labour-force participation rates. In the EU, 25% of women cite care work as their reason for not being in the paid labour force.[41] This compares to 3% of men.

In the UK, women with young children are employed for shorter hours than those without children, while for men it is the other way around.[42] This matches the situation in Mexico where, in 2010, 46% of mothers of very young children were in paid employment compared to 55% of women in households without children.

The figures for men were 99% and 96%, respectively. In the US, female paid employment is actually pretty high amongst younger women, but it sharply declines after motherhood, 'which is being progressively delayed'.[43]

The failure to collect data on women's unpaid workload can also stymie development efforts. Mayra Buvinic, senior fellow at the UN Foundation, points to a history of initiatives in low-income countries littered with training programmes that have failed because they 'have been built on the mistaken assumption that women have plenty of free time, backed by limited data on women's time-intensive work schedules'.[44] Women may sign up for these programmes, but if the initiatives don't account for women's childcare demands, women don't complete them. And that's development money down the drain – and more women's economic potential wasted. In fact, the best job-creation programme could simply be the introduction of universal childcare in every country in the world.

Of course, it's not just childcare that affects female paid employment. Elder care also takes up significant amounts of women's time, and demand is set to increase.[45] Between 2013 and 2050, the global population aged sixty or over is projected to more than double.[46] By 2020, for the first time in history, the number of people aged sixty and over will outnumber children younger than five.[47] And along with getting older, the world is getting sicker. By 2014, nearly a quarter of the world's disease burden was in people aged over sixty – most of it chronic.[48] By 2030 an estimated 6 million older people in the UK (nearly 9% of the total population) will be living with a long-term illness.[49] The EU has already passed this milestone: 10% of its population[50] (around 50 million citizens[51]) are estimated to suffer from two or more chronic conditions. Most of them are sixty-five years and over.[52] In the US, 80% of over-sixty-fives have at least one chronic condition, and 50% have at least two.[53]

All these care needs (the US has an unpaid labour force of 40 million providing care for sick and elderly relatives[54]) affect women's ability to work. Female carers are almost seven times more likely than men to cut back from full-time to part-time work.[55] US women aged between fifty-five and sixty-seven who care for their parents unpaid reduce their paid work hours by, on average, 41%,[56] and 10% of US women caring for someone with dementia have lost job benefits.[57] In the UK, 18% of women who care for someone with dementia have taken a leave of absence from work, and nearly 19% have had to quit work either to become a carer or because their care-giving duties became a priority, while 20% of female carers have gone from working full-time to part-time. This is the case for only 3% of male carers.[58]

If governments want to tap the GDP source of women's increased participation in paid labour it's clear that they have to reduce women's unpaid work: McKinsey found that a decrease in the time British women spend doing unpaid work from five to three hours correlated with a 10% increase in their paid labour-force participation.[59] As we've seen, introducing properly paid maternity and paternity leave is an important step to achieving this, by increasing female paid employment and potentially even helping to close the gender pay gap[60] – which is in itself a boon to GDP. The Institute for Women's Policy Research has found that if women had been paid equally in 2016, the US economy would have produced $512.6 billion more in income – which is 2.8% of 2016's GDP, and represents 'approximately 16 times what the federal and state governments spent in fiscal year 2015 on Temporary Assistance to Needy Families'.[61]

A more dramatic government intervention than the introduction of paid parental leave would be to invest in social infrastructure. The term infrastructure is generally understood to mean the physical structures that underpin the functioning of a modern society: roads, railways, water pipes, power supplies. It doesn't tend to include the public services that similarly underpin the functioning of a modern society like child and elder care.

The Women's Budget Group argues that it should.[62] Because, like physical infrastructure, what the WBG calls social infrastructure 'yields returns to the economy and society well into the future in the form of a better educated, healthier and better cared for population'. Arguably then, this exclusion of care services from the general concept of 'infrastructure' is just another unquestioned male bias in how we structure our economy.

Take early childhood education (ECE) and high-quality formal childcare including for very young toddlers and infants. Investment in these can actually reduce overall education spend because it lowers the level of investment required in remedial education.[63] It also improves cognitive development, educational achievement and health outcomes[64] for children (particularly socio-economically disadvantaged children).[65] All of which increases productivity in the long run.[66]

A report on two ECE pilot studies found that by the age of forty, US children who received ECE were more likely to be employed (76% versus 62%) and to have higher median annual earnings ($20,800 versus $15,300).[67] They were also more likely to own homes (37% versus 28%); a car (82% versus 60%); and to have savings accounts (76% versus 50%). ECE was also found to have wider indirect effects of a lower crime rate, resulting in lower law-enforcement costs. The report concluded that investing in ECE had a greater positive impact on long-term economic growth than business subsidies, and would lead to an extra 3.5% growth in GDP by 2080.

But despite all these potential gains, social-infrastructure investment is often overlooked, in no small part because of the data gap when it comes to unpaid work. This gender data gap has led, Nancy Folbre explains, to its 'pay-off' being 'understated'.[68] In fact, the pay-off could be huge. In the UK it would generate up to 1.5 million jobs, compared to 750,000 for an equivalent investment in construction. In the US, an investment of 2% of GDP in the caring industries 'would create nearly 13 million new jobs, compared to the 7.5 million jobs that would be created by investing 2% of GDP in the construction sector'.[69] And, because the care sector is (currently) a female-dominated industry, many of these new jobs would go to women – remember that increasing female employment drives GDP.

The WBG found that investing 2% of GDP in public care services in the UK, US, Germany and Australia 'would create almost as many jobs for men as investing in construction industries [...] but would create up to four times as many jobs for women'.[70] In the US, where two-thirds of newly created care jobs would go to women compared to only one-third of newly created construction-sector jobs,[71] this investment would increase women's employment rate by up to eight points, reducing the gender employment gap by half.[72] In the UK the investment would reduce the gender employment gap by a quarter (a correction not to be sniffed at given it is women's jobs that have been hardest hit by austerity policies).[73]

As well as increasing female paid employment (and therefore GDP) by actively creating new jobs for women, investing in social infrastructure can also increase female paid employment by reducing the amount of unpaid labour women have to do. The employment rate of UK mothers with children aged three to five is 6% lower than the OECD average. In 2014, 41% of mothers of children under four were employed full-time, compared to 82% of childless women and 84% of fathers.[74] This sex disparity is partly due to societal expectations (enshrined in law via unequal maternity- and paternity-leave allowances) that the mother be the primary carer. But it's also because of the gender pay gap: for many heterosexual couples it makes financial sense for the woman to be the one to reduce her working hours, because she tends to be the one who is earning less.

And then there's the cost of childcare. Recent research from the UK's Department for Education found that 54% of mothers who don't work outside the home said they would like to 'if they could obtain convenient, reliable, and affordable childcare'.[75] But on the whole, they can't. Childcare costs in the UK have outstripped general inflation over the last ten to fifteen years,[76] with UK parents spending 33% of their net household income on childcare against an OECD average of 13%.[77] Unsurprisingly, therefore, the UK has highly unequal take-up of childcare by socio-economic levels, particularly compared to other OECD countries.[78] And this also has a knock-on effect on female paid employment: 29% (this rose to nearly 50% of low- to middle-income mothers) of British women told McKinsey that 'returning to work after having a child is not financially viable – twice the number of men who say the same thing'.[79]

It was a similar story in New York which, in 2012, was found by Pew Research Center to be the most expensive state in the US for childcare.[80] The Center for American Progress found that before the city's mayor introduced universal preschool 'more than one-third of New York families waitlisted for childcare assistance lost their jobs or were unable to work'. In Los Angeles, where preschools face steep funding cuts, an estimated 6,000 mothers are set to give up about 1.5 million work hours, costing an annual total of $24.9 million in lost wages.

There is an easy fix to this problem. One study found that, with consistent childcare, mothers are twice as likely to keep their jobs. Another found that 'government-funded preschool programs could increase the employment rate of mothers by 10 percent'.[81] In

1997, the government of Quebec provided a natural experiment when they introduced a subsidy for childcare services. Following the introduction of the subsidy, childcare prices fell. By 2002 the paid-employment rate of mothers with at least one child aged 1–5 years had increased by 8% and their work hours had increased by 231 per year.[82] Since then, several other studies have found that the public provision of childcare services is 'strongly associated' with higher rates of women's paid employment.[83]

Transferring childcare from a mainly unpaid feminised and invisible form of labour to the formal paid workplace is a virtuous circle: an increase of 300,000 more women with children under five working full-time would raise an estimated additional £1.5 billion in tax.[84] The WBG estimates that the increased tax revenue (together with the reduced spending on social security benefits) would recoup between 95% and 89% of the annual childcare investment.[85]

This is likely to be a conservative estimate, because it's based on current wages – and like properly paid paternity leave, publicly funded childcare has also been shown to lower the gender pay gap. In Denmark where all children are entitled to a full-time childcare place from the age of twenty-six weeks to six years, the gender wage gap in 2012 was around 7%, and had been falling for years. In the US, where childcare is not publicly provided until age five in most places, the pay gap in 2012 was almost double this and has stalled.[86]

We like to think that the unpaid work women do is just about individual women caring for their individual family members to their own individual benefit. It isn't. Women's unpaid work is work that society depends on, and it is work from which society as a whole benefits. When the government cuts public services that we all pay for with our taxes, demand for those services doesn't suddenly cease. The work is simply transferred onto women, with all the attendant negative impacts on female paid labour-participation rates, and GDP. And so the unpaid work that women do isn't simply a matter of 'choice'. It is built into the system we have created – and it could just as easily be built out of it. We just need the will to start collecting the data, and then designing our economy around reality rather than a male-biased confection.

CHAPTER 13

From Purse to Wallet

It was 11 p.m. on the evening of the UK's 2017 general election. The polls had been closed for one hour, and a rumour had started doing the rounds on social media. Youth turnout had gone up. A lot. People were pretty excited about it. 'My contacts are telling me that the turnout from 18–24 year olds will be around 72/73%! Finally the Youth have turnedddd out!! #GE2017' tweeted[1] Alex Cairns, CEO and founder of The Youth Vote – a campaign to engage young people in UK politics. A couple of hours later, Malia Bouattia, then president of the National Union of Students, put out the same statistic in a tweet that went on to be retweeted over 7,000 times.[2] The following morning David Lammy, Labour MP for the London borough of Tottenham, tweeted his congratulations: '72% turnout for 18–25 year olds. Big up yourselves #GE2017'.[3] His tweet received over 29,000 retweets and over 49,000 likes.

There was just one problem: no one seemed to have the data to back any of this up. Not that this stopped news outlets from repeating the claims, all citing either unverified tweets or each other as sources.[4] By Christmas Oxford English Dictionaries had named 'youthquake' as its word of the year, citing the moment 'young voters almost carried the Labour Party to an unlikely victory'.[5] We were witnessing the birth of a zombie stat.

A zombie stat is a spurious statistic that just won't die – in part because it feels intuitively right. In the case of the UK's 2017 general election we needed an explanation for why, contrary to nearly all polling predictions, the Labour Party did so well. An unprecedented increase in youth turnout fitted the bill: Labour had courted the youth vote, the story went, and it had almost won. But then, in January 2018, new data emerged from the British Electoral Survey.[6] There was some debate over how definitive the data was,[7] but the famous youthquake was downgraded to more of a youth-tremor at best. By March no one credible was talking about a 'youth surge' without substantial caveats, and the 72% statistic was firmly on life support.[8]

The British youthquake that never was had a fairly short life for a zombie stat. This is partly because while secret ballots preclude the possibility of absolutely conclusive polling data, we do at least collect data on them. A lot of data, in fact: elections are hardly an underresearched topic. But when a zombie stat emerges in an area where data is scarce, the stat becomes much harder to explode.

Take the claim that '70% of those living in poverty are women.' No one is quite sure where this statistic originated, but it's usually traced to a 1995 UN Human Development Report, which included no citation for the claim.[9] And it pops up everywhere, from newspaper articles, to charity and activist websites and press releases, to statements and reports from official bodies like the ILO and the OECD.[10]

There have been efforts to kill it off. Duncan Green, author of *From Poverty to Power*, brands the statistic 'dodgy'.[11] Jon Greenberg, a staff writer for fact-checking website Politifact, claims, citing World Bank data,[12] that 'the poor are equally divided by gender', with, if anything, men being slightly worse off. Caren Grown, senior director of Gender Global Practice at the World Bank, bluntly declares the claim to be 'false,' explaining that we lack the sex-specific data (not to mention a universally understood definition of what we mean by 'poverty') to be able to say one way or the other.[13]

And this is the problem with all this debunking. The figure may be false. It may also be true. We currently have no way of knowing. The data Greenberg cites no doubt does indicate that poverty is a gender-blind condition, but the surveys he mentions, impressive though their sample size may be ('a compilation of about 600 surveys across 73 countries'), are entirely inadequate to the task of determining the extent of feminised poverty. And having an accurate measure is important, because data determines how resources are allocated. Bad data leads to bad resource allocation. And the data we have at the moment is incredibly bad.

Gendered poverty is currently determined[14] by assessing the relative poverty of households where a man controls the resources (male-headed household) versus households where a woman controls the resources (female-headed household).[15] There are two assumptions being made here. First, that household resources are shared equally between household members, with all household members enjoying the same standard of living. And second, that there is no difference between the sexes when it comes to how they allocate resources within their households. Both assumptions are shaky to say the least.

Let's start with the assumption that all members of a household enjoy an equal standard of living. Measuring poverty by household means that we lack individual level data, but in the late 1970s, the UK government inadvertently created a handy natural experiment that allowed researchers to test the assumption using a proxy measure.[16] Until 1977, child benefit in Britain was mainly credited to the father in the form of a tax reduction on his salary. After 1977 this tax deduction was replaced by a cash payment to the mother, representing a substantial redistribution of income from men to women. If money were shared equally within households, this transfer of income 'from wallet to purse' should have had no impact on how the money was spent. But it did. Using the proxy measure of how much Britain was spending on clothes, the researchers found that following the policy change the country saw 'a substantial increase in spending on women's and children's clothing, relative to men's clothing'.

Of course, 1977 was a long time ago, and you'd be forgiven for hoping things might have changed since then. Unfortunately, however, this is the most recent sex-disaggregated data we have for the UK, so it's impossible to say. But we do have more recent data from other countries (including Ireland, Brazil, the US, France, Bangladesh and the Philippines) and it is not encouraging. Money continues not to be shared equally between couples, and money controlled by women continues to be more likely to be spent on children (a gender-neutral word which itself hides a wealth of inequalities[17]) than money controlled by men.[18] So unless the UK is a secret feminist paradise (I can confirm that it is not), it's safe to say that very little has changed.

This being the case, the British government's decision to introduce a new benefit called universal credit (UC) is unfortunate. UC merges several benefits and tax credits (including child tax credit) and, unlike the benefits it replaces, it is paid by default into the account of the main earner in each household.[19] Given the gender pay gap, this is almost universally the man in heterosexual couples – and 'almost universally' is as exact as we're going to get on this, because the UK's Department for Work and Pensions isn't collecting sex-disaggregated data on who the money is going to. So, in the UK at least, the data gap on gendered poverty is about to get even bigger.

Now we've established that men and women have different spending priorities, it should be clear that there is a big question mark over the second assumption, that living in a male-headed versus a female-headed household has no implications for your standard of living. And this is indeed what the data we have shows. In Rwanda and Malawi, children from female-headed households were healthier than children from male-headed households – even when the male-headed households had higher incomes.[20]

An analysis of the 2010 Karnataka Household Asset Survey in India was even more damning.[21] When merely comparing female-headed to male-headed households, there was not much gender difference found in poverty levels. However, when poverty was assessed on an individual level, the difference was dramatic, with, wait for it, 71% of those living in poverty being women. And within those living in poverty it was women who experienced the greatest level of deprivation. Perhaps most damning for the validity of using household wealth to measure gendered poverty, the majority of poor women belonged to 'non-poor' households.

It's time for us to kill off the zombie assumptions that poverty can be determined at a household level, or that 'female-headed' has the same implications for male poverty that 'male-headed' has for female poverty. They are based on faulty data and non-gender-sensitive analysis. More than this, they add to and perpetuate the gender data gap. And they have led to some policy decisions that are disastrous for women.

In the US, nearly all married couples file a joint tax return. They don't have to: they have the choice of filing either individually or as a couple. But the system incentivises them so strongly – through lower taxes and access to certain tax credits – to file jointly that 96% of married couples do.[22] And the result, in practice, is that most married women in the US get over-taxed on their income.

The US tax system is progressive, which means there are several tax bands. The first $10,000 or so that you earn gets taxed at a lower rate than the next $10,000 you earn, and so on. So, let's say you earn $20,000 and your friend earns $60,000. For the first $20,000 of her income, you and your friend will pay the same amount of tax. But she will pay a higher rate of tax on the income she earns above that. That is, unless you happen to be married to that person and you file a joint tax return with her. In that case, you and your partner are treated as a single economic unit, with an income of $80,000, and how your tax is calculated changes.

In a married couple's joint tax return, the couple must 'stack' their wages. The higher earner (given the gender pay gap this is usually the man) is designated the 'primary earner', and their income occupies the lower tax bracket. The lower earner (usually the woman) becomes the 'secondary earner', and their income occupies the higher tax bracket. To return to our couple earning $60,000 and $20,000, the person earning $20,000 will be taxed on that income as if it is the final $20,000 of an $80,000 salary, rather than all she earns. That is, she will pay a much higher rate of tax on that income than if she filed independently of her higher-earning husband.

Defenders of the married-couple tax return will point out that *overall* the couple is paying less tax by filing together. And this is true. But because, as we've seen, the assumption that household resources are shared equally is flawed to say the least, a *couple* paying less tax doesn't necessarily translate into more money in the secondary earner's pocket than if she'd filed individually. And this is before we even address any issues of how financial abuse may be making the joint filing system even worse for women. In short, the current US tax system for married couples in effect penalises women in paid employment, and in fact several studies have shown that joint filing disincentivises married women from paid work altogether (which, as we have also seen, is bad for GDP).[23]

The US is not alone in having a tax system that, by failing to account for gender, ends up discriminating against women. A recent paper expressed bafflement at how 'many OECD countries' were passing legislation in an attempt to reduce the gender pay gap while at the same time effectively increasing it through their family tax and transfer systems.[24] Two such countries are the UK and Australia where, although married couples file separate income tax returns, most benefits and tax credits still breach the principle of independent taxation.

The UK's Marriage Allowance gives the main wage earner (usually the man) a tax break in couples where the lower earner is on £11,500 or less.[25] This bolsters the gender pay gap on two fronts: supplementing male income, while also creating a perverse incentive for women to work fewer paid hours. Japan has a similarly male-biased married-couples tax break. Since 1961, the 'head of household' (normally a man) has been able to 'claim a tax deduction of ¥380,000 ($3,700) as long as his spouse's income does not exceed ¥1.03m (around $10,000)'. A 2011 survey by Japan's labour ministry found that 'more than a third of married women who worked part time and deliberately curtailed their hours did so to keep the tax deduction'.[26]

In a slightly different example of a hidden gendered bias, Argentina's tax system provides a rebate almost four times higher for employees than for the self-employed. Gender comes into it because men are more likely to be employed in the formal economy, while women are more likely to be self-employed in the informal economy.[27] So the tax system is essentially covertly giving a higher rebate to men than to women.

There's a fairly simple reason why so many tax systems discriminate against women, and that is that we don't systematically collect data on how tax systems affect them. In other words, it's because of the gender data gap. The impact of taxation on women is 'an underdeveloped area of research' according to a 2017 report from the European Parliament, which called for more sex-disaggregated data on the issue.[28] Even countries such as Spain, Finland and Ireland that have taken steps to analyse their budgeting from the perspective of gender, usually focus only on spending, not tax. In the EU, Austria 'is one of the few countries where the government has defined specific goals for the tax system, such as promoting a more equal division of paid and unpaid work between women and men, enhancing the labour participation of women and reducing the gender pay gap'. Meanwhile, a 2016 survey of EU member states found that only Finland and Sweden have strictly individualised income tax systems.[29]

The tax system's woman problem extends beyond the zombie assumption that household resources are allocated equally between the sexes: it encompasses the theory of taxation itself – at least in its current form. Since the 1980s, governments around the world have been less interested in taxes as a means to redistribute resources, seeing tax more as a potential retardant to growth that must be contained. The result has been lower taxes on capital, corporations and high-income earners, and an increase in loopholes and incentives so that multinational corporations and the super-rich can avoid and evade tax. The idea is not to 'distort otherwise efficient market processes'.[30]

When gender has come into this framework at all, it has been solely in the context of how tax might harm growth by disincentivising women to enter paid employment. What isn't considered is how a tax system focused so narrowly on enabling 'growth' benefits men at the expense of women. Cuts in the top rates of income tax disproportionately benefit men because of the gender pay gap. For the same reason, the majority of women in the world are not in a position to make use of the various tax loopholes an expensive accountant can afford you. Decreases in (or non-enforcement of) wealth and asset taxes also disproportionately benefit men, because men are far more likely to control such resources.[31]

But it's not just about benefiting men over women. These male-biased benefits actually come at women's expense, because as we've seen, women have to fill the resulting service gaps with their unpaid care work. In 2017, the Women's Budget Group pointed out that at the same time that austerity measures were having a particularly severe impact on women in the UK, 'tax giveaways disproportionately benefitting men will cost the Treasury £44bn per annum by 2020'.[32] These include a £9 billion cut in fuel and alcohol duties, a £13 billion cut in corporation tax, and a loss of £22 billion from raising income tax and National Insurance thresholds. Together, these tax giveaways accounted for more than the total annual cuts in social security spending – which makes it clear that this isn't a matter of resources, so much as (gendered) spending priorities.

The problem of low tax revenues in low-income countries is exacerbated by cross-border tax-avoidance techniques: multinational companies often 'negotiate tax holidays or incentives as a condition for bringing their business to developing countries', costing developing countries an estimated $138 billion in revenue annually. Well, the argument goes, if massive corporations paying zero taxes while they exploit cheap labour is the only way to get them there ... Only it isn't. The OECD has found that 'such incentives are rarely a primary reason for investment in developing countries'.[33] Women's cheap labour, on the other hand, is certainly quite the draw. Nevertheless, such tax systems are sometimes 'imposed as conditions on developing countries by international financial institutions'.[34]

In a parallel to UK tax giveaways that outpace its spending cuts, the IMF estimates that developing countries lose $212 billion per year from tax-avoidance schemes, which far outstrips the amount they receive in aid.[35] Over a third of the world's total unrecorded offshore financial wealth is thought to be secretly held in Switzerland, which recently faced questions from the UN 'over the toll that its tax and financial secrecy policies take on women's rights across the globe'.[36] A 2016 analysis by the Center for Economic and Social Rights (CESR) found that the amount of money lost to tax dodging by multinational copper firms such as the Swiss-headquartered Glencore in Zambia, could finance 60% of the country's health budget. CESR also estimated that the Indian government lost out on up to '$1.2 billion in direct tax revenue from the funds held in just one bank branch in Switzerland – comparable to as much as 44% of [India's] expenditure on women's rights, and 6% of total social spending in the country in 2016'.[37]

Governments need money, so they have to make up these losses somehow. Many of them turn to consumption taxes because they are easy to collect and difficult to evade. Low-income countries raise 'about two-thirds of their tax revenue through indirect taxes such as VAT, and just over a quarter through income taxes'.[38] A recent International Labour Organization analysis found that 138 governments (ninety-three developing and forty-five developed countries) are planning to either increase and/or extend consumption taxes, primarily through VAT.[39]

This increase disproportionately affects women too. Not just because they are over-represented among the poor (the poorer you are the higher a proportion of your income goes on consumption), but also because they tend to bear the responsibility of buying food and household goods. And because women's paid labour supply is more elastic (in no small part because of the gender pay gap), increasing VAT can have the effect of pushing women to spend more time in unpaid work in order to produce in the household what they might otherwise buy on the market.

This problem is exacerbated by an often gender-insensitive allocation of what products do and don't have VAT added, driven by an overall lack of research based on sex-disaggregated data on the impact specific consumption tax rates and exemptions have.[40] VAT is not generally added to products that are seen as 'essential', so in the UK, food is exempt because it's considered essential, while iPhones are not because they are not. But one man's frivolity is another woman's essential, and around the world women have been campaigning to get male-dominated legislators to recognise that sanitary products are not luxury items. In some countries they've even succeeded.

It's clear that tax systems around the world, presented as the objective trickle-down of market-driven forces have intensely gendered impacts. They have been created based on non-sex-disaggregated data, and male-default thinking. Together with our woman-blind approach to GDP and public spending, global tax systems are not simply failing to alleviate gendered poverty: they are driving it. And if the world cares about ending inequality, we need to adopt an evidence-based economic analysis as a matter of urgency.

CHAPTER 14

Women's Rights are Human Rights

What the past two chapters have shown is that there are substantial gender data gaps in government thinking, and the result is that governments produce male-biased policy that is harming women. These data gaps are in part a result of failing to collect data, but they are also in part a result of the male dominance of governments around the world. And while we may not think of male-dominated government as a gender data gap problem, the evidence makes it clear that female perspective matters.

Several US studies from the 1980s to the 2000s have found that women are more likely to make women's issues a priority and more likely to sponsor women's issues bills.[1] In the UK, a recent analysis of the impact female MPs have had in Westminster since 1945 found that women are more likely to speak about women's issues, as well as family policy, education and care.[2] An analysis[3] of the impact of female representation across nineteen OECD countries[4] between 1960 and 2005 also found that female politicians are more likely to address issues that affect women.

The OECD study also found that women's words translated into action. As female political representation increased in Greece, Portugal and Switzerland, these countries experienced an increase in educational investment. Conversely, as the proportion of female legislators in Ireland, Italy and Norway decreased in the late 1990s, those countries experienced 'a comparable drop in educational expenditures as a percentage of GDP'. As little as a single percentage point rise in female legislators was found to increase the ratio of educational expenditure. Similarly, a 2004 Indian study of local councils in West Bengal and Rajasthan found that reserving one-third of the seats for women increased investment in infrastructure related to women's needs.[5] A 2007 paper looking at female representation in India between 1967 and 2001 also found that a 10% increase in female political representation resulted in a 6% increase in 'the probability that an individual attains primary education in an urban area'.[6]

In short, decades of evidence demonstrate that the presence of women in politics makes a tangible difference to the laws that get passed. And in that case, maybe, just maybe, when Bernie Sanders said, 'It is not good enough for someone to say, "I'm a woman! Vote for me!" ', he was wrong. The problem isn't that anyone thinks that's good enough. The problem is that no one does. On the other hand plenty of people seem to think that a candidate being a woman is a good enough reason *not* to vote for her. Shortly before the 2016 US presidential election, the *Atlantic* published the results of a focus group of undecided voters.[7] The main takeaway was that Hillary Clinton was just too ambitious.

This is not a groundbreaking opinion. From Anne Applebaum ('Hillary Clinton's extraordinary, irrational, overwhelming ambition'[8]), to Hollywood mogul, democratic donor and 'one-time

Clinton ally'[9] David Geffen ('God knows, is there anybody more ambitious than Hillary Clinton?'[10]), via Colin Powell ('unbridled ambition'[11]), Bernie Sanders' campaign manager ('don't destroy the Democratic Party to satisfy the secretary's ambitions'[12]), and, of course, good old Julian Assange ('eaten alive by her ambitions'[13]), the one thing we all seem to be able to agree on (rare in this polarised age) is that Hillary Clinton's ambition is unseemly. Indeed, so widespread is this trope it earned itself a piece in the *Onion* headlined, 'Hillary Clinton is too Ambitious to be the First Female President'.[14]

Being the first woman to occupy the most powerful role in the world does take an extraordinary level of ambition. But you could also argue that it's fairly ambitious for a failed businessman and TV celebrity who has no prior political experience to run for the top political job in the world – and yet ambition is not a dirty word when it comes to Trump.

Associate professor of psychology at UC Berkeley Rodolfo Mendoza-Denton has a cognitive explanation for why we may view Clinton's ambition as 'pathological'.[15] She 'was forging into a territory that is overwhelmingly associated in people's minds with men'. As a result, he explains, voters experienced her candidacy as a norm violation. And norm violations are, Mendoza-Denton writes, 'quite simply, aversive, and are often associated with strong negative emotion'.

There's a very simple reason that a powerful woman is experienced as a norm violation: it's because of the gender data gap. I personally grew up heavily buying into the myth that women are just . . . a bit rubbish. Yes, this was partly because that's how women are represented in the media (consumerist, trivial, irrational) but it's also because women are so *under*-represented. I was one of those girls being taught, via a curriculum, a news media and a popular culture that were almost entirely devoid of women, that brilliance didn't belong to me. I wasn't being shown women I could look up to (either past or present). I wasn't being taught about female politicians, female activists, female writers, artists, lawyers, CEOs. All the people I was taught to admire were men, and so in my head power, influence, and ambition equated with maleness. And, if I'm being really honest, I think I experienced this norm violation as well. I was all too ready to accept the idea that female bosses were just too ambitious – which as we all know is code for bitch.

The unpalatable truth is that it is still considered unladylike for a woman to want to be president. A 2010 study found that both male and female politicians are seen as power-seeking, but that this is only a problem for female politicians.[16] In a similar vein, Mendoza-Denton conducted a study which found that context determines how 'assertive' men and women are judged to be.[17] In a stereotypically 'male' context (car mechanic, Wall Street, president of the United States) a woman is judged to behave more assertively than a man saying exactly the same as her. And while it was OK if a bit odd for men to be assertive in a 'female' context (choosing curtains, planning a child's birthday party), it was definitely not OK for a woman to be assertive in any context. Assertive women are bossy.

The social downer on women being seen to seek professional power is partly because social power (being seen as warm and caring) is women's 'consolation prize for renouncing competition with men,' write psychology professors Susan Fiske and Mina Cikara.[18] Social power for women is therefore intrinsically incompatible with professional power: if a woman wants to be seen as competent she has to give up being seen as warm.

But so what. So you're disliked. So you're seen as cold. Suck it up. If you don't like the heat, get back to the kitchen, right?

Wrong. That would be to assume that men face the same heat for being seen as cold. They don't. The 2010 study didn't just find that female politicians were seen as less caring. It found that this perception inspired moral outrage in both male and female study participants, who viewed such women with contempt, anger and/or disgust. This was not the case for their male counterparts. Molly Crockett, associate professor of experimental psychology at Oxford University, has an explanation for this disparity: being seen as uncaring is a norm violation for women in a way that it just isn't for men. 'There is an expectation', she tells me, 'that on average women are going to be more pro-social than men.' Any deviation by a woman from what is seen (no matter how illogically) as a 'moral' stance therefore shocks us more.

Given the clear significance of gender when it comes to these issues, you would hope that this might be an area of research that bucks the gender-data-gap trend. It does not. Imagine my excitement when I came across a paper published in January 2017 entitled 'Faced with exclusion: Perceived facial warmth and competence influence moral judgments of social exclusion'.[19] Given the findings of Fiske and Cikara about women's warmth/competence trade-off this should have been an extremely useful paper. As the authors explain, 'people's moral judgment about social exclusion can be influenced by facial appearance, which has many implications in intergroup research'. That is, people's decisions about whether or not it's fair that someone is being ostracised or bullied can be influenced by what the victim looks like.

Indeed. Unfortunately, the study authors 'used male faces only for reasons of test efficiency', making the study absolutely worthless when it comes to the group most affected by this issue, i.e. women. Fiske and Cikara explain that gender, 'is a salient, and perhaps the most salient, social category', with gender stereotyping often being immediate and unconscious: 'the mere sight of a woman can immediately elicit a specific set of associated traits and attributions, depending on the context'. Still, at least the test was efficient.

'It's actually kind of shocking how little attention there's been to gender in the morality literature,' says Crockett. But on the other hand, maybe it's not: the study of morality, Crockett tells me, is 'really aiming at trying to uncover human universals'. At the point she mentions 'universals', of course, male-default-thinking alarm bells start ringing in my head. Many academics in the field of morality subscribe to 'very egalitarian, utilitarian, impartial views of what is right', Crockett continues, and they perhaps impose those norms 'onto the research that we do'. The alarm bells ring off the hook.

But the next thing she says provides something of an explanation for how male-default thinking could be so prevalent in a world that is, after all, 50% female. 'It's just a feature of human psychology,' she explains, to assume that our own experiences mirror those of human beings in general. This is a concept in social psychology that is sometimes called 'naive realism' and sometimes called 'projection bias'. Essentially, people tend to assume that our own way of thinking about or doing things is typical. That it's just normal. For white men this bias is surely magnified by a culture that reflects their experience back to them, thereby making it seem even more typical. Projection bias amplified by a form of confirmation bias, if you like. Which goes some way towards explaining why it is so common to find male bias masquerading as gender neutrality. If the majority of people in power are men – and they are – the majority of people in power just don't see it. Male bias just looks like common sense to them. But 'common sense' is in fact a product of the gender data gap.

Mistaking male bias for impartial, universal, common sense means that when people (men) come across someone trying to level the playing field, it's often all they can see (perhaps because they read it as bias). A 2017 paper found that while white male leaders are praised for promoting diversity, female and ethnic minority leaders are penalised for it.[20] This is partly because by promoting diversity, women and ethnic minorities remind white men that these female ethnic-minority leaders are, in fact, women and ethnic minorities. And so all the stereotypes that go along with that become salient: bossy, assertive, cold and all the rest. Conversely, ethnic minority and female leaders 'avoid negative stereotypes when they engage in low levels of diversity-valuing behavior'. At last, empirical proof for what most women (even if they don't admit it to themselves) have always known, at least implicitly: playing along with patriarchy is of short-term, individual benefit to a woman. There's just the minor issue of being on borrowed time.

The finding that engaging in 'diversity-valuing behavior' reminds people that a woman is in fact a woman perhaps explains how Sanders came to think that all Clinton said was 'vote for me, I'm a woman' – because the data shows that she certainly didn't. A word-frequency analysis of her speeches by Vox journalist David Roberts revealed that Clinton 'mostly talked about workers, jobs, education and the economy, exactly the things she was berated for neglecting. She mentioned jobs almost 600 times, racism, women's rights and abortion a few dozen times each.' But, pointed out US writer Rebecca Solnit in her *London Review of Books* piece on the election, 'she was assumed to be talking about her gender all the time, though it was everyone else who couldn't shut up about it'.[21]

What all of this means on a grander scale is that democracy is not a level playing field: it is biased against electing women. This is a problem, because male and female legislators inevitably bring different perspectives to politics. Women lead different lives to men because of both their sex and their gender. They are treated differently. They experience the world differently, and this leads to different needs and different priorities. Like a male-dominated product-development team, a male-dominated legislature will therefore suffer from a gender data gap that will lead it to serve its female citizens inadequately. And most of the world's governments are male-dominated.

As of December 2017, women made up an average of 23.5% of the world's parliamentarians, although this figure hides significant regional variation: Nordic parliaments are on average 41.4% female while Arab parliaments are on average 18.3% female.[22] Women account for 10% or less of parliamentarians in thirty-one countries, including four countries that have no female parliamentarians at all. And in most countries precious little is being done to remedy this.

In 2017 the UK's Women and Equalities Committee produced a report with six recommendations for the government to increase female representation in Parliament.[23] They were all rejected.[24] One of the recommendations was for the government to allow all-women shortlists (AWS) in local as well as general elections, and to extend their legality beyond the current 2030 cut-off point. In the British system, each political party holds an internal election for every constituency to decide which candidate will stand for them in a general election. AWS are used in these internal elections if a party wants to ensure that their general-election candidate will be a woman.

AWS were first used in the UK's 1997 elections. In January 1997, the United Kingdom tied with St Vincent & the Grenadines and Angola in the world rankings of female parliamentarians.[25] With a 9.5% female House of Commons, they all sat in joint fiftieth place. But by December of the same year the UK had suddenly shot up to twentieth place, because in May it had an election. And in that election the Labour Party, the UK's main opposition party, made use of AWS for the first time. The effect was dramatic. The number of female Labour MPs leapt from thirty-seven to 101 (the overall rise in female MPs was from sixty to 120).

In the 2017 UK general election, Labour used AWS for 50% of its winnable seats, and 41% of the candidates the party fielded were female. The Tories and the Lib Dems, neither of whom used AWS, fielded 29% each. The UK's House of Commons is currently (2018) 32% female, which places it at thirty-ninth in the world rankings – a drop in standing which is partly a result of other countries catching up, and partly a result of the dominance of the Conservative Party which still doesn't use AWS (43% of Labour's MPs are female compared to 21% of the Conservatives).

It is clear that Labour's use of AWS has driven a significant proportion of the increase in female MPs. The government's refusal to extend their legality beyond 2030 is therefore tantamount to legislating for a resumption of male bias in British democracy. Perhaps they haven't read the data on the impact female politicians bring to legislation. Or, perhaps they have.

The British government's refusal to extend AWS to local elections is if anything more perplexing, because female representation is even worse in local government. Britain's trend towards devolution was meant to be about giving power back to local communities (local government, on which Britain spends £94 billion each year, plays a vital role in providing services that women in particular depend on). But the evidence uncovered by a 2017 report commissioned by women's charity the Fawcett Society suggests that it is mainly giving power back to men.[26]

The Fawcett Society report found that nine councils across England and Wales still have all-male cabinets, while only 33% of council chief executives are female. Just one in three councillors in England is a woman, up only five percentage points in two decades. All six of the newly elected metro mayors are men (in the latest Liverpool election none of the main parties even fielded a female candidate), and just 12% of cabinet members in devolved areas are women.

The Fawcett report is all the evidence we have, because the data is not being collected by government, so unless this particular charity continues to collect the data, it will be impossible to monitor progress. And yet the government's reasoning for refusing to extend AWS to local or mayoral elections is that the 'evidence base is as yet underdeveloped'.[27] Given they also refused the committee's most basic recommendation to force parties to collect and publish candidate-diversity data (on the basis of the 'regulatory burden which this would impose'), this stance leaves those who would like to see a less male-biased form of democracy take hold in Britain at something of a disadvantage.

Three of the recommendations in the Women and Equalities report concerned the implementation of quotas, and it was not surprising that these were rejected: British governments have traditionally been opposed to such measures, seeing them as anti-democratic. But evidence from around the world shows that political gender quotas don't lead to the monstrous regiment of incompetent women.[28] In fact, in line with the LSE study on workplace quotas, studies on political quotas have found that if anything, they 'increase the competence of the political class in general'. This being the case, gender quotas are nothing more than a corrective to a hidden male bias, and it is the current system that is anti-democratic.

The form of quota that is available to a country depends on the electoral system it operates. In the UK, each of the country's 650 constituencies has a single MP. This MP is voted in using 'first past the post' (FPTP), which means that the candidate with the most votes gets returned to Parliament. Since there is only one candidate per constituency, in a FPTP system all-women shortlists are really the only practicable corrective to male bias.

In Sweden, a party list is used. In this system, each constituency is represented by a group of MPs allocated under proportional representation (PR). Every party draws up a list of candidates per constituency, with candidates set down in order of preference. The more votes a party receives, the more candidates from its list are elected to represent that constituency. The lower a candidate is listed, the less likely she is to win a seat.

In 1971, only 14% of Swedish parliamentarians were female.[29] The Social Democratic Party (SDP) decided to try to address this discrepancy, first with a recommendation in 1972 that party districts should place 'more women' on electoral lists.[30] By 1978, this had evolved into a recommendation that lists reflect the proportion of female party members, and in 1987 a 40% minimum target was introduced. None of these measures had a significant effect on the number of female MPs elected: you could have a 50% female list, but if all the women were down at the bottom, they weren't likely to win a seat.

So in 1993 the SDP introduced what is known as a 'zipper' quota. Two lists must be produced: one of male candidates and one of female candidates. These two lists are then 'zipped' together, so you end up with a list that alternates male and female candidates. In the 1994 election that followed, female representation leapt eight points,[31] and has never dipped below 40% since[32] (although the proportion of women in parliament has been slipping as Sweden has increasingly been voting for more right-wing parties that don't operate gender quotas).

Compare this to South Korea, which provides an instructive example of how something as seemingly unrelated to gender as an electoral system can in fact make all the difference to female representation. South Korea operates a mixed electoral system with around 18% of its seats allocated under PR,[33] and the rest working in the same way as the UK Parliament: single-member districts (SMD) elected under a FPTP. Both systems operate under a quota for female representation.

When the PR system quotas were increased from 30% to 50% for the 2004 elections, female representation more than doubled in the South Korean parliament. This sounds impressive, but they were starting from a low base, because while the parties more or less stick to the quota in the PR system, it's a different story in the SMD. Here, 30% of candidates are supposed to be women, but in a recent election women comprised only 7% of the Saenuri Party's and 10% of the Democratic United Party's SMD candidates. If both SMD and PR quotas were adhered to, the South Korean parliament would be around 33.6% female. As it is, female representation currently stands at 15.7%.

It's not hard to see why there's such a stark difference in quota compliance between the two systems: FPTP and SMD electoral systems are a zero-sum game.[34] Winner takes all. And so while on a macro level all-women shortlists in such systems are a fair corrective to an unfair system, on a micro level they certainly feel less fair – particularly to the specific man who wasn't even allowed to compete.

This was the argument of two rejected Labour candidates, Peter Jepson and Roger Dyas-Elliott. In 1996, the two men brought a legal challenge against the Labour Party in the UK, arguing that AWS fell foul of the 1975 Sex Discrimination Act. Given what we know about the invisible positive discrimination that operates in favour of men this was perhaps not in the spirit of the Act. It was, however, in the letter of it, and Jepson and Dyas-Elliott won their case. AWS were briefly outlawed before being brought back in via the Labour government's 2002 Act. Originally intended to run until 2015, in 2008 Harriet Harman, then Labour's deputy leader, announced that its run would be extended to 2030.[35] Meanwhile, Dyas-Elliot was most recently to be found in court receiving a restraining order for sending a rival MP's wife a dead bird.[36]

Worldwide, the countries with the highest levels of female political representation tend to use PR.[37] With this in mind, and given South Korea's and Sweden's experiences, perhaps the UK's Women and Equalities Committee shouldn't have called for quotas as a first step. If they really want to see female representation increase in Parliament, perhaps their first demand should be full electoral reform. But increasing female representation is only half the battle, because it's not much use getting women elected if they're prevented from doing their job effectively once they're there. And frequently, they are.

Clare Castillejo, a specialist in fragile states, writes that women's influence in government is often limited by their exclusion from male-dominated patronage networks.[38] Women may be present at formal talks, but this isn't much good if the men are forming back-room quid pro quo networks (something Castillejo cautions is particularly common in post-conflict settings[39]) and going off to have the real discussion in 'informal spaces that women cannot access'.[40]

The practice of excluding women from decision-making is widespread, and it is one of the most efficient ways (second only to not electing women at all) that this male-biased system has of siphoning off gendered data in the form of female life experience and perspective. In a 2011 survey of US legislators, 40% of women disagreed with the statement 'The leaders in my legislature are as likely to consult with the women in the legislature as the men when making important decisions' (interestingly, only 17% of men disagreed with it).[41] Similarly, a 2017 report on local government in the UK referenced 'informal networks within local government where real power lies' and in which women are 'less likely to be involved'.[42]

But male politicians don't have to escape to all-male safe spaces to sideline women. There are a variety of manoeuvres they can and do employ to undercut their female colleagues in mixed-gender settings. Interrupting is one: 'females are the more interrupted gender,' concluded a 2015 study that found that men were on average more than twice as likely to interrupt women as women were to interrupt men.[43] During a televised ninety-minute debate in the run-up to the 2016 US presidential election, Donald Trump interrupted Hillary Clinton fifty-one times, while she interrupted him seventeen times.[44] And it wasn't just Trump: journalist Matt Lauer (since sacked after multiple allegations of sexual harassment[45]) was also found to have interrupted Clinton more often than he interrupted Trump. He also 'questioned her statements more often',[46] although Clinton was found to be the most honest candidate running in the 2018 election.[47]

Patronising women is another manoeuvre, an infamous example being then British prime minister David Cameron's 'Calm down, dear' to Labour MP Angela Eagle in 2011.[48] In the Inter-Parliamentary Union's (IPU) 2016 global study on sexism, violence and harassment against female politicians, one MP from a European parliament said 'if a woman speaks loudly in parliament she is "shushed" with a finger to the lips, as one does with children. That never happens when a man speaks loudly'.[49] Another noted that she is 'constantly asked – even by male colleagues in my own party – if what I want to say is very important, if I could refrain from taking the floor.' Some tactics are more brazen. Afghan MP Fawzia Koofi told the *Guardian* that male colleagues use intimidation to frighten female MPs into silence – and when that fails, 'The leadership cuts our microphones off'.[50]

Highlighting the hidden gender angle of having a single person (most often a man) in charge of speaking time in parliament, one MP from a country in sub-Saharan Africa (the report only specified regions so the women could remain anonymous) told the IPU that the Speaker had pressured one of her female colleagues for sex. Following her refusal, 'he had never again given her the floor in parliament'. It doesn't necessarily even take a sexual snub for a Speaker to refuse women the floor: 'During my first term in parliament, parliamentary authorities always referred to statements by men and gave priority to men when giving the floor to speakers,' explained one MP from a country in Asia.

The IPU report concluded that sexism, harassment and violence against female politicians was a 'phenomenon that knew no boundaries and exists to different degrees in every country'. The report found that 66% of female parliamentarians were regularly subjected to misogynistic remarks from their male colleagues, ranging from the degrading ('you would be even better in a porn movie') to the threatening ('she needs to be raped so that she knows what foreigners do').

Political abuse is a distinctly gendered phenomenon.[51] During the 2016 Democratic primaries, Hillary Clinton received almost twice as many abusive tweets as Bernie Sanders. The most common word associated with her was 'bitch'. Bitch was also the most common term used in tweets about Australian ex-PM Julia Gillard, who between 2010 and 2014 was similarly the target of almost twice as many abusive messages as her political rival Kevin Rudd. One European MP told the IPU that she once received more than 500 rape threats on Twitter over a period of four days.[52] Another woman had been sent information about her son – 'his age, the school he attends, his class, etc. – threatening to kidnap him'.

Sometimes it's not 'just' threats. More than one in five female parliamentarians surveyed by the IPU had been 'subjected to one or more acts of sexual violence', while a third had witnessed sexual violence being committed against a female colleague. During the 2010 elections in Afghanistan, nearly all of the female candidates received threatening phone calls,[53] and some female MPs in the country require round-the-clock protection.[54] 'Almost every day I fear for my life,' Afghan MP Fawzia Koofi told the *Guardian* in 2014;[55] a year later one of her female colleagues died in a car bomb – the second deadly attack on a female politician in Afghanistan in the space of three months.[56]

The aggression seems to increase along with the proportion of female politicians. Research from around the world (including saintly Scandinavia) has shown that as female representation increases, so does hostility against female politicians.[57] Especially from their male colleagues. Studies[58] in the US and New Zealand have shown that men 'become more verbally aggressive and controlling of both committee hearings and parliamentary debates following an expansion in the proportion of women in the legislature'. Another study found that as the proportion of women in US Congress increases (bear in mind Congress is only 19.4% female[59]), women are less likely to achieve leadership positions within their parties.[60] Further research[61] from the US and Argentina has shown that having large numbers of female legislators is 'tied to both women's diminished success in passing legislation and reduced chances of being appointed to "masculine" and "powerful" committees'.[62] In a similar vein, US analysis has found that framing human rights issues as women's rights issues makes male politicians less likely to support legislation, and if a rights bill is mainly sponsored by women, it ends up being watered down and states are less likely to invest resources.[63] It seems that democracy – in so far as it pertains to women – is broken.

Working in the context of such extreme psychological warfare inevitably affects women's ability to do their jobs. Many women told the IPU that they had restricted their travel, made sure they went home before nightfall, or only travelled when accompanied.[64] Others self-censor, particularly when it comes to speaking up about women's issues[65] (which tend to generate the most aggression[66]), some going as far as dispensing with social media altogether, and in this way deprive themselves 'of a forum in which to disseminate and debate their ideas'.

Others simply stand down. Violence against female politicians in Asia and Latin America has been shown to make them less likely to stand for re-election and more likely to leave after fewer terms compared to male politicians.[67] 'I don't know if I will be a candidate in the next elections,' one Asian MP told the IPU, 'because I need to think about not causing too much harm to my family.'[68] Meanwhile one in three female politicians in Swedish local politics 'reportedly considered giving up their positions as a result of threatening incidents'.[69]

The abuse faced by female politicians also makes women more reluctant to stand in the first place. More than 75% of British women on a programme for aspiring female leaders said that sexist abuse of female politicians online 'was a point of concern when considering whether to pursue a role in public life'.[70] In Australia, 60% of women aged eighteen to twenty-one and 80% of women over thirty-one said the way female politicians were treated by the media made them less likely to run for office.[71] Nigeria experienced a 'marked decline' in the number of female politicians elected to the country's congress between 2011 and 2015; a study by the US NGO the National Democratic Institute found that this could be 'attributed to the violence and harassment that women in office face'.[72] And, as we have seen, this decline in female representation will give rise to a gender data gap that in turn will result in the passing of less legislation that addresses women's needs.

The evidence is clear: politics as it is practised today is not a female-friendly environment. This means that while *technically* the playing field is level, in reality women operate at a disadvantage compared to men. This is what comes of devising systems without accounting for gender.

Sheryl Sandberg's approach for navigating hostile work environments, outlined in her book *Lean In*, is for women to buckle up and push through. And of course that is part of the solution. I am not a female politician, but as a woman with a public profile I get my own share of threats and abuse. And, unpopular as this opinion may be,

I believe that the onus is on those of us who feel able to weather the storm, to do so. The threats come from a place of fear. In fact, a gender-data-gap-driven fear: certain men, who have grown up in a culture saturated by male voices and male faces, fear what they see as women taking away power and public space that is rightfully theirs. This fear will not dissipate until we fill in that cultural gender data gap, and, as a consequence, men no longer grow up seeing the public sphere as their rightful domain. So, to a certain extent, it is an ordeal that our generation of women needs to go through in order that the women who come after us don't.

This is not to say that there are no structural solutions. Take the issue of women being interrupted. An analysis of fifteen years of Supreme Court oral arguments found that 'men interrupt more than women, and they particularly interrupt women more than they interrupt other men'.[73] This goes for male lawyers (female lawyers weren't found to interrupt at all) as well as judges, even though lawyers are meant to stop speaking when a justice starts speaking. And, as in the political sphere, the problem seems to have got worse as female representation on the bench has increased.

An individualist solution might be to tell women to interrupt right back[74] – perhaps working on their 'polite interrupting'[75] skills. But there's a problem with this apparently gender-neutral approach, which is that it isn't gender-neutral in effect: interrupting simply isn't viewed the same way when women do it. In June 2017 US Senator Kamala Harris was asking an evasive Attorney General Jeff Sessions some tough questions. When he prevaricated once too often, she interrupted him and pressed him to answer. She was then in turn (on two separate occasions) interrupted and admonished by Senator John McCain for her questioning style.[76] He did not do the same to her colleague Senator Rob Wyden, who subjected Sessions to similarly dogged questioning, and it was only Harris who was later dubbed 'hysterical'.[77]

The problem isn't that women are irrationally polite. It's that they know – whether consciously or not – that 'polite' interrupting simply doesn't exist for them. So telling women to behave more like men – as if male behaviour is a gender-neutral human default – is unhelpful, and in fact potentially damaging. What is instead called for is a political and work environment that accounts both for the fact that men interrupt more than women do, and that women are penalised if they behave in a similar way.

It has become fashionable for modern workplaces to relax what are often seen as outmoded relics of a less egalitarian age: out with stuffy hierarchies, in with flat organisational structures. But the problem with the absence of a formal hierarchy is that it doesn't actually result in an absence of a hierarchy altogether. It just means that the unspoken, implicit, profoundly non-egalitarian structure reasserts itself, with white men at the top and the rest of us fighting for a piece of the small space left for everyone else. Group-discussion approaches like brainstorming, explains female leadership trainer Gayna Williams, are 'well known to be loaded with challenges for diverse representation', because already-dominant voices dominate.[78]

But simple adjustments like monitoring interruptions[79] and more formally allocating a set amount of time for each person to speak have both been shown to attenuate male dominance of debates. This is in fact what Glen Mazarra, a showrunner at FX TV drama *The Shield*, did when he noticed that female writers weren't speaking up in the writer's room – or that when they did, they were interrupted and their ideas overtaken. He instituted a no-interruption policy while writers (male or female) were pitching. It worked – and, he says, 'it made the entire team more effective'.[80]

A more ambitious route would be changing the structure of governance altogether: away from majority-based, and towards unanimous decision-making. This has been shown to boost women's speech participation and to mitigate against their minority position[81] (a 2012 US study found that women only participate at an equal rate in discussions when they are in 'a large majority'[82] – interestingly while individual women speak less when they are in the minority, individual men speak the same amount no matter what the gender proportion of the group).

Some countries have attempted to legislate against the more extreme ways in which women's voices are shut out from power. Since 2012, Bolivia has made political violence against a woman elected to or holding public office a criminal offence; in 2016 they also passed a law preventing anyone with a background of violence against women from running for political office.

But on the whole, most countries proceed as if female politicians do not operate at a systemic disadvantage. While most parliaments have codes of conduct, these are generally focused on maintaining a gender-neutral 'decorum'. Most countries have no official procedure for settling sexual-harassment complaints, and it's often up to whoever happens to be in charge (usually a man) to decide whether sexism is in fact indecorous and therefore against rules. Often they don't. One female MP told the IPU that when she demanded a point of order following a sexist insult from a colleague, the Speaker had rejected her motion. 'I cannot control what another member thinks of you,' she was told.

The UK used to have a gender-specific code of conduct for local government, overseen by an independent body which had the power to suspend councillors. But this was discarded under the 2010 coalition government's 'Red Tape Challenge'. It is now up to each local authority to decide which standards to set and how to enforce them. The government's recommendations for how this should be done included only one vague reference to promoting 'high standards of conduct' and did not mention non-discrimination at all.[83] There is no longer any clear mechanism by which councillors can be suspended for non-criminal misconduct.[84]

It is unsurprising then that by 2017, when the Fawcett Society produced a report on local government, the women's charity found 'a harmful culture of sexism in parts of local government politics which would not be out of place in the 1970s', where 'sexism is tolerated, and viewed as part of political life', and where almost four in ten women councillors have had sexist remarks directed at them by other councillors.[85] One female councillor described 'a culture of demeaning younger women and dismissing the contribution that women make'. A women's group was described as 'the wives club'; a dinner with a senior national political speaker 'was promoted as an opportunity for 'the wives' to dress up'. When she and a fellow female colleague challenged the behaviour they were described as 'aggressive', and 'referred to by demeaning, sexist nicknames'. Her emailed questions have gone ignored; she has been excluded from meeting notifications; and she described her contributions to discussions as 'tolerated rather than welcomed'. On social media her own party colleagues told her to 'run away little girl and let the grown-ups do their job'.

There are two central points to take away from this section. The first is that when you exclude half the population from a role in governing itself, you create a gender data gap at the very top. We have to understand that when it comes to government the 'best' doesn't have to mean 'those who have the money, the time and the unearned confidence from going to the right school and university'. The best when it comes to government means the best as a whole, as a working group. And in that context, the best means diversity. Everything we've seen so far in this book shows us without a doubt that perspective does matter. The data accrued from a lifetime of being a woman matters. And this data belongs at the very heart of government.

Which leads to the second point: the data we already have makes it abundantly clear that female politicians are not operating on a level playing field. The system is skewed towards electing men, which means that the system is skewed towards perpetuating the gender data gap in global leadership, with all the attendant negative repercussions for half the world's population. We have to stop wilfully closing our eyes to the positive discrimination that currently works in favour of men. We have to stop acting as if theoretical, legal equality of opportunity is the same as true equality of opportunity. And we have to implement an evidence-based electoral system that is designed to ensure that a diverse group of people is in the room when it comes to deciding on the laws that govern us all.

PART VI

When it Goes Wrong

CHAPTER 15

Who Will Rebuild?

When Hillary Clinton wanted to speak about women's rights at the 1995 United Nations Fourth World Conference on Women in Beijing, even her own side was dubious.[1] 'People were saying: "This is a not an important issue for the US government, it's a nice thing and I'm glad you care about it, but if the First Lady of the United States goes and actually speaks about women's rights, that elevates an issue that in the midst of everything else going on – the collapse of the USSR and the transition of the former Soviet states and Warsaw Pact nations and Rwanda and Bosnia, there was so much else going on in the world – maybe you should speak about it from afar."' As we will see (and as the US administration already knew at the time) what was 'going on' in Rwanda and Bosnia was the mass and systematic rape of women.

When things go wrong – war, natural disaster, pandemic – all the usual data gaps we have seen everywhere from urban planning to medical care are magnified and multiplied. But it's more insidious than the usual problem of simply forgetting to include women. Because if we are reticent to include women's perspectives and address women's needs when things are going well, there's something about the context of disaster, of chaos, of social breakdown, that makes old prejudices seem more justified. And we're always ready with an excuse. We need to focus on rebuilding the economy (as we've seen, this is based on a false premise). We need to focus on saving lives (as we *will* see this is also based on false premise). But the truth is, these excuses won't wash. The real reason we exclude women is because we see the rights of 50% of the population as a minority interest.

The failure to include women in post-disaster efforts can end in farce. 'They built houses without any kitchens,' Maureen Fordham, a professor of disaster resilience, tells me. It was 2001, and an earthquake had just hit Gujarat, a state in western India. Thousands of people died and nearly 400,000 homes were destroyed. So new homes were needed, but Gujarat's rebuilding project had a major data gap: women weren't included or even consulted in the planning process. Hence the kitchenless homes. In some confusion I ask Fordham how people were expected to cook. 'Well, quite,' she replies, adding that the homes were also often missing 'a separate area that's usually attached to a house where the animals are kept', because animal care isn't on the whole a male responsibility. 'That's women's work.'

If this sounds like an extreme one-off, it isn't. The same thing happened in Sri Lanka four years later.[2] It was after the 2004 Boxing Day tsunami which swept across the coasts of fourteen countries bordering the Indian Ocean, killing a quarter of a million people in its wake. And just like in Gujarat, Sri Lanka's rebuilding programme didn't include women, and, as a result, they built homes without kitchens. A related issue arises in refugee camps when humanitarian agencies distribute food that must be cooked – but forget to provide cooking fuel.[3]

The US has a similar history of forgetting about women in post-disaster relief efforts. Fordham tells me about the redevelopment scheme set up in Miami following 1992's Hurricane Andrew. 'They called it "We Will Rebuild".' The problem was, the 'we' who were planning the rebuilding were nearly all men: of the fifty-six people on the decision-making board (reportedly an 'invitation-only group of Miami insiders'[4]) only eleven were women.

This male-dominated 'we' were criticised at the time as 'an uptown group trying to deal with a downtown problem'. One woman simply saw 'the good ole boy network once more taking charge, running things when they had no real idea of what the problems were, especially the problems of women. It was business as usual.' And what this good ole boy network wanted to rebuild was business centres, the skyscrapers, the Chamber of Commerce facility, at a time when 'thousands were still suffering from [a] lack of basic necessities [and] community services'. They completely missed, says Fordham, 'things like nursery schools or health centres', as well as the smaller-scale informal workplaces, which, as we've seen, are particularly relevant to women's needs. In Miami, disgruntled women's rights activists set up 'Women Will Rebuild' to address the gaps in the official scheme.

We Will Rebuild was a while ago now, but when Hurricane Katrina hit New Orleans thirteen years later, it became clear that lessons had not been learned. Over 30,000 people were displaced by the August 2005 hurricane (at the time, the US was in the top ten of countries with 'major internally displaced populations of concern'[5]) and the single largest category of these internally displaced populations were African American women. But despite their dominance amongst those affected, African American women's voices were barely heard at all in planning efforts, either before or after the storm hit.[6] This omission constituted a major gender data gap and resulted in a failure to direct resources towards those who were most vulnerable, which, said a 2015 Institute of Women's Policy Research (IWPR) report, could easily have been predicted with proper research. Instead, by failing to consult women about their needs, planners were responsible for what the IWPR called a 'third disaster' following the twin disasters of the hurricane and subsequent flooding. And this third disaster was, 'like the failure of the levees, of human origin'.

Most former tenants of New Orleans public housing wanted to – and assumed they would – return to their former homes after the clean-up. After all, 'the Bricks', as the four large housing projects within the city of New Orleans are known, were still standing. More than this, according to the US Department of Housing and Urban Development, they were structurally sound and would be habitable after cleaning. But it was not to be. Even as 'affordable and structurally sound homes in New Orleans remained in high demand', funding was announced for the buildings to be demolished. They would be replaced with mixed-income housing which included only 706 public housing units compared to the 4,534 that had existed before.

Like 'We Will Rebuild' in Miami before them, the planners seemed to place business interests above the needs of 'the now permanently displaced thousands of individuals, all low-income and the majority black women'. In their legal response to a 2007 lawsuit, the Housing Authority of New Orleans claimed that they had surveyed former tenants and the majority had said that they did not want to return to New Orleans. This is the opposite of what IWPR found, leaving many with the suspicion that 'the decision to destroy the buildings may have been less about repairing disaster damage, or responding to the needs of those who had suffered losses and experienced traumas, and more about opportunistic urban redevelopment'.

Residents wanted to return to the Bricks because, like Brazil's *favelas*, these public housing projects provided more than shelter: they provided a social infrastructure, filling in the gaps left by a laissez-faire state. 'Public housing might not have been the best, but everybody was somebody's momma back up in there,' one woman told IWPR. When the women were displaced and dispersed – and then had their homes demolished – they lost all that. But because we don't measure women's unpaid work, a need to maintain such informal ties once again was not factored into any rebuilding efforts. The social networks provided by the housing projects also meant that women felt safer, which in turn made them more mobile. '[T]he city wasn't too bad,' explained one woman, 'because everybody knew everybody, and then once you got towards Orleans and Claiborne [streets], you were safe because you knew everyone.'

The mobility of women living in the Bricks was also supported by the regular buses and variety of stores in walking distance. But again, all of that has changed. Walking is no longer an option for many of the displaced women who now live miles from the nearest stores. And bus schedules have been changed: where buses used to come along every fifteen minutes, it's now not unusual to have to wait for an hour. One woman lost her job as a result. Rather like the planners behind Brazil's Minha Casa, Minha Vida, transporting low-income women to their places of employment does not seem to have been considered a priority by the architects of New Orleans' regeneration.

There is no international law requiring that women's voices be included in post-disaster planning – although based on the evidence perhaps there should be. When it comes to post-conflict contexts, however, we have UN Security Council Resolution 1325.

UNSCR 1325 'urges all actors to increase the participation of women and incorporate gender perspectives in all United Nations peace and security efforts'. Following 'decades of lobbying' from women's rights activists,[7] this landmark resolution was passed in 2000. But eighteen years on, further progress has been minimal. For a start, the available data is scant[8] – which in itself is suggestive about the seriousness with which this resolution is taken. As for what data does exist, it's hardly encouraging. Only two women have ever served as chief negotiators and only one woman has ever signed a final peace accord as chief negotiator.[9] Funding for the implementation of policies related to women's rights in post-conflict contexts remains 'inadequate',[10] as does progress on the basic requirement of including women in all delegations.[11] Even where women are included, they remain in the minority and excluded from positions of power, and in some areas, we have even regressed: in 2016 only half of that year's signed peace agreements contained gender-specific provisions, compared to 70% of signed peace agreements in 2015. In the June 2017 Afghanistan peace talks, women made up 6% of negotiators, 0% of mediators, and 0% of signatories.

Causal data for the sudden reversal between 2016 and 2017 is not available, but a clue comes courtesy of a participant at an off-the-record round table on women, peace and security at the International Peace Institute in New York in 2014. 'The UN and other powerbrokers succumb to requests not to have women in the room,' the participant claimed. 'When the local government says "We don't want women," the international community compromises and says "OK." '[12] As in post-disaster contexts, the reasoning given varies (cultural sensitivities; including women would delay the negotiations; women can be included after an agreement has been reached) but they all boil down to the same line that's been used to fob women off for centuries: we'll get to you after the revolution.

It's a rationale that is clearly a function of sexism, a symptom of a world that believes women's lives are less important than 'human' lives, where 'human' means male. But the ease with which international agencies toss UNSCR 1325 out of the window is not just sexist. It's foolish. The presence of women at the negotiating table not only makes it more likely that an agreement will be reached,[13] it also makes it more likely that the peace will last. Analysis of 182 peace agreements signed between 1989 and 2011 demonstrated that when women are included in peace processes there is a 20% increase in the probability of an agreement lasting at least two years, and a 35% increase in the probability of an agreement lasting at least fifteen years.[14]

This isn't necessarily a matter of women being better at negotiating: it's at least in part what they negotiate *for*. Clare Castillejo, the specialist in governance and rights in fragile states, points out that 'women frequently bring important issues to the peace-building agenda that male elites tend to overlook', such as the inclusivity and accessibility of processes and institutions and the importance of local and informal spheres.[15] In other words, as ever, the presence of women fills in a data gap – and an important one: recent quantitive data analysis has found 'compelling evidence' that countries where women are kept out of positions of power and treated as second-class citizens are less likely to be peaceful.[16] In other words: closing the gender data gap really is better for everyone.

CHAPTER 16

It's Not the Disaster that Kills You

The irony of excluding women's voices when it all goes wrong is that it is exactly in these extreme contexts that old prejudices are *least* justified, because women are already disproportionately affected by conflict, pandemic and natural disaster. The data on the impact of conflict (mortality, morbidity, forcible displacement) on women is extremely limited and sex-disaggregated data is even rarer. But the data we do have suggests that women are disproportionately affected by armed conflict.[1] In modern warfare it is civilians, rather than combatants, who are most likely to be killed.[2] And while men and women suffer from the same trauma, forcible displacement, injury and death, women also suffer from female-specific injustices.

Domestic violence against women increases when conflict breaks out. In fact, it is more prevalent than conflict-related sexual violence.[3] To put this in context, an estimated 60,000 women were raped in the three-year Bosnian conflict and up to 250,000 in the hundred-day Rwandan genocide. UN agencies estimate that more than 60,000 women were raped during the civil war in Sierra Leone (1991–2002); more than 40,000 in Liberia (1989–2003); and at least 200,000 in the Democratic Republic of the Congo since 1998.[4] Because of data gaps (apart from anything else, there is often no one for women to report to), the real figures in all these conflicts are likely to have been much higher.

In the breakdown of social order that follows war, women are also more severely affected than men. Levels of rape and domestic violence remain extremely high in so-called post-conflict settings, 'as demobilized fighters primed to use force confront transformed gender roles at home or the frustrations of unemployment'.[5] Before the 1994 genocide in Rwanda, the average age for marriage for a girl was between twenty and twenty-five years; in the refugee camps during and after the genocide, the average age for marriage was fifteen years.[6]

Women are also more likely than men to die from the indirect effects of war. More than half of the world's maternal deaths occur in conflict-affected and fragile states, and the ten worst-performing countries on maternal mortality are all either conflict or post-conflict countries. Here, maternal mortality is on average 2.5 times higher, and this is partly because post conflict and disaster relief efforts too often forget to account for women's specific healthcare needs.

For over twenty years, the Inter-agency Working Group on Reproductive Health in Crises has called for women in war zones or disaster areas to be provided with birth kits, contraception, obstetrics care and counselling. But, reports the *New York Times*, 'over the past two decades, that help has been delivered sporadically, if at all'.[7] One report found that pregnant women are left without obstetrical care, 'and may miscarry or deliver under extremely unsanitary conditions.'

This can also be an issue in post-disaster zones: following the Philippines' 2013 typhoon in which 4 million people were left homeless, an estimated 1,000 women were giving birth every day, with almost 150 of them expected to experience life-threatening conditions.[8] Birthing facilities and equipment had been destroyed by the typhoon, and women were dying.[9] But when the United Nations Population Fund asked donor nations for funds to pay for hygiene kits, staff at temporary maternity wards and counselling for rape victims, the response was 'lukewarm', with only 10% of the amount needed being raised.[10]

Post-conflict and post-disaster zones are also particularly vulnerable to the spread of infectious diseases – and women die in greater numbers than men when pandemics hit.[11] Take Sierra Leone, the country at the heart of the 2014 Ebola outbreak, and which has the highest maternal mortality rate in the world: 1,360 mothers die per every 100,00 live births (for comparison, the OECD average is fourteen per 100,000[12]), and one in seventeen mothers have a lifetime risk of death associated to childbirth.[13] The government has recently released data revealing that at least 240 pregnant women die every month in Sierra Leone.[14]

Throw Ebola into the mix and women suddenly had two types of death to fear: from childbirth and from Ebola. In fact it was worse than that, because pregnant women were at increased risk of contracting Ebola due to their high levels of contact with health services and workers:[15] the *Washington Post* reported that two of the three largest outbreaks of Ebola 'involved transmission of the virus in maternity settings'.[16] The fact that Ebola decimated healthcare workers (themselves mainly women) made the feminised risk even higher: the *Lancet* estimated that in the three countries affected by the virus, an extra 4,022 women would die every year as a result of the shortage.[17]

The reluctance to factor gender into relief efforts is partly due to the still-persistent attitude that since infectious diseases affect both men and women, it's best to focus on control and treatment 'and to leave it to others to address social problems that may exist in society, such as gender inequalities after an outbreak has ended'.[18] Academics are also at fault here: a recent analysis of 29 million papers in over 15,000 peer-reviewed titles published around the time of the Zika and Ebola epidemics found that less than 1% explored the gendered impact of the outbreaks.[19] But, explains a WHO report, the belief that gender doesn't matter is a dangerous position which can hinder preventative and containment efforts, as well as leaving important insights into how diseases spread undetected.[20]

Failing to account for gender during the 2009 H1N1 (swine flu virus) outbreaks meant that 'government officials tended to deal with men because they were thought to be the owners of farms, despite the fact that women often did the majority of work with animals on backyard farms'.[21] During the 2014 Ebola outbreak in Sierra Leone, 'initial quarantine plans ensured that women received food supplies, but did not account for water or fuel'. In Sierra Leone and other developing countries, fetching fuel and water is the job of women (and of course fuel and water are necessities of life), so until the plans were adjusted, 'women continued to leave their houses to fetch firewood, which drove a risk of spreading infection'.[22]

Women's care-taking responsibilities also have more deadly consequences for women in pandemics. Women do the majority of care for the sick at home. They also make up the majority of 'traditional birth attendants, nurses and the cleaners and laundry workers in hospitals, where there is risk of exposure', particularly given these kinds of workers 'do not get the same support and protection as doctors, who are predominantly men'.[23] Women are also those who prepare a body for a funeral, and traditional funeral rites lead many to be infected.[24] In Liberia, during the 2014 Ebola epidemic, women were estimated to make up 75% of those who died from the disease;[25] in Sierra Leone, the 'epicentre' of the outbreak, UNICEF estimated that up to 60% of those who died were women.[26]

A 2016 paper[27] also found that in the recent Ebola and Zika epidemics international health advice did not 'take into account women's limited capacity to protect themselves from infection'.[28] In both cases, advice issued was based on the (inaccurate) premise that women have the economic, social or regulatory power 'to exercise the autonomy contained in international advice'. The result was that already-existing gender inequalities were 'further compounded' by international health advice.

We need to address the gender data gap when it comes to post-disaster relief with some urgency, because there is little doubt that climate change is making our world more dangerous. According to the World Meteorological Organisation, it's nearly five times more dangerous than it was forty years ago: between 2000 and 2010 there were 3,496 natural disasters from floods, storms, droughts and heat waves, compared to 743 natural disasters in the 1970s.[29] And beyond analyses that suggest climate change can be a factor in the outbreak of conflict[30] and pandemic,[31] climate change itself is causing deaths. A 2017 report in the journal *Lancet Planetary Health* predicted that weather-related disasters will cause 152,000 deaths a year in Europe between 2071 and 2100.[32] This compares to 3,000 deaths a year between 1981 and 2010.[33] And, as we will see, women tend to dominate the figures of those who die in natural disasters as well.

We didn't have firm data on the sex disparity in natural-disaster mortality until 2007, when the first systematic, quantitative analysis was published.[34] This examination of the data from 141 countries between 1981 to 2002 revealed that women are considerably more likely to die than men in natural disasters, and that the greater the number of people killed relative to population size, the greater the sex disparity in life expectancy. Significantly, the higher the socio-economic status of women in a country, the lower the sex gap in deaths.

It's not the disaster that kills them, explains Maureen Fordham. It's gender – and a society that fails to account for how it restricts women's lives. Indian men have been found to be more likely to survive earthquakes that hit at night 'because they would sleep outside and on rooftops during warm nights, a behavior impossible for most women'.[35] In Sri Lanka, swimming and tree climbing are 'predominantly' taught to men and boys; as a result, when the December 2004 tsunami hit (which killed up to four times as many women as men[36]) they were better able to survive the floodwaters.[37] There is also a social prejudice against women learning to swim in Bangladesh, 'drastically' reducing their chances of surviving flooding,[38] and this socially created vulnerability is compounded by women not being allowed to leave their home without a male relative.[39] As a result, when cyclones hit, women lose precious evacuation time waiting for a male relative to come and take them to a safe place.

They also lose time waiting for a man to come and tell them there's a cyclone coming in the first place. Cyclone warnings are broadcast in public spaces like the market, or in the mosque, explains Fordham. But women don't go to these public spaces. 'They're at home. So they're totally reliant on a male coming back to tell them they need to evacuate.' Many women simply never get the message.

A male-biased warning system is far from the only part of Bangladesh's cyclone infrastructure that has been built without reference to women's needs. Cyclone shelters have been built 'by men for men', says Fordham, and as a result they are often far from safe spaces for women. Things are slowly changing, but there is a 'huge legacy' of old-style cyclone shelters, which are basically just 'a *very* large concrete box'. Traditionally the shelter is just one big mixed-sex space. There are usually no separate latrines for men and women: 'just a bucket in the corner and you might have 1,000 people in these places sheltering'.

Beyond the obvious problem of a single bucket for 1,000 people, the lack of sex segregation essentially locks women out of the shelters. 'It's embedded in Bangladeshi culture that women cannot mix with men and boys outside of their family males,' explains Fordham, for fear of bringing shame on the family. Any woman mixing with those males 'is just fair game for any kind of sexual harassment and worse. So the women won't go to the shelters.' The result is that women die at much higher rates (following the 1991 cyclone and flood the death rate was almost five times as high for women as for men[40]) simply for want of sex-segregated provision.

On the subject of the violence women face in disaster contexts, we know that violence against women increases in the 'chaos and social breakdown that accompany natural disaster' – but, in part because of that self-same chaos and social breakdown, we don't know by exactly how much. During Hurricane Katrina local rape crisis centres had to close, meaning that in the days that followed no one was counting or confirming the number of women who had been raped.[41] Domestic-violence shelters also had to close, with the same result. Meanwhile, as in Bangladesh, women were experiencing sexual violence in gender-neutral storm shelters. Thousands of people who had been unable to evacuate New Orleans before Katrina hit were temporarily housed in Louisiana's Superdome. It didn't take long for lurid stories of violence, of rapes and beatings, to start circulating. There were reports of women being battered by their partners.[42]

'You could hear people screaming and hollering for people to help them, "Please don't do this to me, please somebody help me"', one woman recalled in an interview with IWPR.[43] 'They said things didn't happen at the Superdome. They happened. They happened. People were getting raped. You could hear people, women, screaming. Because there's no lights, so it's dark, you know.' She added, 'I guess they was just grabbing people, doing whatever they wanted to do.' Precise data on what happened to whom in Hurricane Katrina has never been collated.

For women who try to escape from war and disaster, the gender-neutral nightmare often continues in the refugee camps of the world. 'We have learned from so many mistakes in the past that women are at a greater risk for sexual assault and violence if they don't have separate bathrooms,' says Gauri van Gulik, Amnesty International's deputy director for Europe and Central Asia.[44] In fact international guidelines state that toilets in refugee camps should be sex-segregated, marked and lockable.[45] But these requirements are often not enforced.

A 2017 study by Muslim Women's charity Global One found that 98% of female refugees in Lebanon did not have access to separate latrines.[46] Research by the Women's Refugee Commission has found that women and girls in accommodation centres in Germany and Sweden are vulnerable to rape, assault and other violence because of a failure to provide separate latrines, shower facilities or sleeping quarters. Mixed living and sleeping quarters can mean women develop skin rashes from having to keep their hijab on for weeks.

Female refugees regularly[47] complain that the remote location[48] of many toilets is worsened by a lack of adequate lighting both on the routes to the latrines and in the facilities themselves. Large areas of the infamous Idomeni camp in Greece were described as 'pitch-black' at night. And although two studies have found that installing solar lighting or handing out individual solar lights to women in camps has had a dramatic impact on their sense of safety, it's a solution that has not been widely adopted.[49]

So most women find their own solutions. A year after the 2004 tsunami women and girls in Indian displacement camps were still walking in pairs to and from the community toilet and bathing facilities to ward off harassment from men.[50] A group of Yezidi women who ended up in Nea Kavala camp in northern Greece after fleeing sexual slavery under ISIS formed protection circles so they could accompany each other to the toilet. Others (69% in one 2016 study[51]), including pregnant women who need frequent toilet trips, simply don't go at night. Some women in reception centres in Germany have resorted to not eating and drinking, a solution also reported by female refugees in Idomeni, at the time Greece's largest informal refugee camp.[52] According to a 2018 *Guardian* report, some women have taken to wearing adult nappies.[53]

Some of the failure to protect women from male violence in European camps can be put down to the speed with which authorities in, for example, Germany and Sweden (who to their credit have taken far more refugees than most), have had to respond to the crisis.[54] But this is not the whole story, because women in detention centres around the world experience the same problems with male guards. Women at a US immigration facility in 2005 reported that guards used a camera phone to take pictures of them while they were sleeping, as well as when they came out of the showers and bathrooms.[55] In 2008, a seventeen-year-old Somali refugee detained at a Kenyan police station was raped by two policemen when she left her cell to use the toilet.[56] Yarl's Wood Detention centre in the UK has been dogged for years by multiple cases of sexual abuse and assault.[57]

Given the steady stream of abuse reports from around the world, perhaps it's time to recognise that the assumption that male staff can work in female facilities as they do in male facilities is another example of where gender neutrality turns into gender discrimination. Perhaps sex-segregation needs to extend beyond sanitation facilities, and perhaps no male staff should be in positions of power over vulnerable women. Perhaps. But if this is going to happen, authorities would first have to countenance the idea that male officials might be exploiting the women they are meant to be variously helping, guarding or processing. And, currently, authorities are not countenancing this.

In an email to humanitarian news agency IRIN a spokesperson for the Regional Office of Refugee Affairs in Berlin (LAF) wrote that 'After countless conversations with shelter managers, I can assure you that there is no unusual occurrence [of sexualised violence] reported from emergency or community shelters.'[58] Despite multiple accounts of sexual harassment and abuse they were, they said, 'confident there is no significant problem'. Similarly, news website *BuzzFeed* reports that in Europe the possibility that male border guards might trade sex for entry is all but denied.[59] And yet a 2017 *Guardian* report revealed that 'Sexual violence and abuse was widespread and systematic at crossings and checkpoints. A third of the women and children interviewed said their assailants wore uniforms or appeared to be associated with the military.'[60]

The LAF substantiated their claim of 'no significant problem' by pointing to the 'very low numbers of police reports', with only ten cases of 'crimes against the sexual freedom of a person' involving women living in refugee shelters registered by Berlin police in the whole of 2016.[61] But are police statistics a reliable measure of the problem, or is this yet another gender data gap? When *BuzzFeed* reporters contacted the national police of the major European transit countries (Greece, Macedonia, Serbia, Croatia and Hungary) for any information they had about gender-based violence, many simply did not respond to 'repeated requests for information'. The Hungarian national police did reply, but only to say that 'it does not collect information related to asylum-seekers, including reports of rape or attempted sexual assault'. The Croatians said they 'could not disaggregate crime reports by victim category', although in any case they 'had no reports of asylum-seekers experiencing gender-based violence'. This may of course be true, although not because it's not happening. Several women's organisations who work with refugees point out that although many of the women they work with have been groped and harassed at shelters, a mixture of cultural and language barriers mean that a 'very, very high number of sexually motivated attacks go unreported'.[62]

The data gap when it comes to sexual abuse is compounded in crisis settings by powerful men who blur the lines between aid and sexual assault, exploiting their position by forcing women to have sex with them in order to receive their food rations.[63] The data gaps here are endemic, but the evidence we do have suggests that this is a common scenario in post-disaster environments,[64] and one which has recently hit headlines worldwide, as first Oxfam and then various other international aid agencies were rocked by allegations of sexual abuse by their workers, and subsequent cover-ups.[65]

The irony of ignoring the potential for male violence when it comes to designing systems for female refugees is that male violence is often the reason women are refugees in the first place.[66] We tend to think of people being displaced because of war and disaster: this is usually why men flee. But this perception is another example of male-default thinking: while women do seek refuge on this basis, female homelessness is more usually driven by the violence women face from men. Women flee from 'corrective' rape (where men rape a lesbian to 'turn her straight'), from institutionalised rape (as happened in Bosnia), from forced marriage, child marriage and domestic violence. Male violence is often why women flee their homes in low-income countries, and it's why women flee their homes in the affluent West.

Homelessness has historically been seen as a male phenomenon, but there is reason to doubt the official data on this issue. Joanne Bretherton, research fellow at the University of York's Centre for Housing Policy, explains that women are actually 'far more likely to experience homelessness than men',[67] while in Australia the 'archetypal homeless person' is now 'a young women aged 25–34, often with a child, and, increasingly, escaping violence'.[68] But this 'serious social problem'[69] has been grossly underestimated – and it's a gender data gap that is in many ways a product of how researchers define and measure homelessness.[70] According to the Canadian Centre for Policy Alternatives (CCPA) 'much of the research on homelessness [...] lacks a comprehensive gender-based analysis'.[71]

Homelessness is usually measured by counting those who use homeless services, but this approach only works if men and women are equally likely to use these services, and they aren't. Women made homeless as a result of domestic violence are often likely to seek refuge in domestic-violence shelters rather than homeless shelters. In the UK this means that they will not be counted as homeless.[72] They are also likely to live in precarious arrangements with other people, 'without their own front door, privacy and their own living space, and without access to any housing of their own to which they have a legal right'.[73] Sometimes, as witnessed by the recent rise in 'sex for rent' agreements across the UK, they will, like women in refugee camps, be sexually exploited.[74]

According to Canadian research, women fall into these precarious arrangements because they don't feel safe in the official emergency accommodation, especially when it's mixed sex.[75] And these safety issues are not a product of women's imaginations: the CCPA calls the levels of violence experienced by women in shelters 'staggering'. Supposedly 'gender-neutral' services that are 'presumed to be equally accessible for men and women', concludes the CCPA, 'actually put women at significant risk'.

Female homelessness is therefore not simply a result of violence: it is a lead predictor of a woman experiencing violence.[76] Women in the US are choosing to live rough rather than access shelters they perceive as dangerous.[77] Katharine Sacks-Jones, director of women-at-risk charity Agenda, explains that in the UK homelessness services are 'often set up with men in mind', and that they 'can be frightening places for vulnerable women who've experienced abuse and violence'.[78]

Gender-sensitive provision is not just about safety, however, it's also about health. In the UK, homeless shelters can (and do) request free condoms from the NHS,[79] but they cannot request free menstrual products. As a result, shelters can only provide menstrual products for free if they happen to have spare funds (unlikely) or if they receive a donation. In 2015, a campaign group called The Homeless Period petitioned the UK government to fund the provision of menstrual products as they do condoms.[80] Despite questions being raised in Parliament, government funding has not been forthcoming, although in March 2017 the campaign group announced a partnership with Bodyform to donate 200,000 packs of sanitary products by 2020.[81] Campaigners in America have been more successful: in 2016 New York City became the first US city to provide free tampons and pads in public schools, homeless shelters and correctional facilities.[82]

Female refugees have also not been spared the impact of the chronic global failure to account for the fact that women menstruate. Funding for this essential resource is often not forthcoming,[83] and the result is that women and girls can go for years without access to menstrual products.[84] Even where hygiene kits are distributed, they have traditionally been 'designed for household-level distribution with no adjustment for the number of menstruating females in each household'.[85] Distribution is also too often designed without regard for the cultural taboo around menstruation: expecting women to feel able to request menstrual products from male workers or in front of male family members;[86] and not providing culturally sensitive products or disposal methods.[87]

These gaps in provision affect women's health and freedom. Reduced to resorting to unhygienic substitutes ('old rags, pieces of moss, pieces of mattress[88]), one study found that over 50% of women had 'suffered from urinary-tract infections which were often left untreated'.[89] And 'because of the stigma surrounding menstruation and the risk of leakages', women are restricted in their movements, unable to 'access food, get services, information, interact with other people'.

Closing the gender data gap will not magically fix all the problems faced by women, whether or not they are displaced. That would require a wholesale restructuring of society and an end to male violence. But getting to grips with the reality that gender-neutral does not automatically mean gender-equal would be an important start. And the existence of sex-disaggregated data would certainly make it much harder to keep insisting, in the face of all the evidence to the contrary, that women's needs can safely be ignored in pursuit of a greater good.

Afterword

The quarrels of popes and kings, with wars and pestilences, in every page; the men so good for nothing and hardly any women at all – it is very tiresome.

Jane Austen

It took about two hours for Daina Taimina to find the solution that had eluded mathematicians for over a century. It was 1997, and the Latvian mathematician was participating in a geometry workshop at Cornell University. David Henderson, the professor leading the workshop, was modelling a hyperbolic plane constructed out of thin, circular strips of paper taped together. 'It was disgusting,' laughed Taimina in an interview.[1]

A hyperbolic plane is 'the geometric opposite' of a sphere, explains Henderson in an interview with arts and culture magazine *Cabinet*.[2] 'On a sphere, the surface curves in on itself and is closed. A hyperbolic plane is a surface in which the space curves away from itself at every point.' It exists in nature in ruffled lettuce leaves, in coral leaf, in sea slugs, in cancer cells. Hyperbolic geometry is used by statisticians when they work with multidimensional data, by Pixar animators when they want to simulate realistic cloth, by auto-industry engineers to design aerodynamic cars, by acoustic engineers to design concert halls. It's the foundation of the theory of relativity, and 'thus the closest thing we have to an understanding of the shape of the universe'.[3] In short, hyperbolic space is a pretty big deal.

But for thousands of years, hyperbolic space didn't exist. At least it didn't according to mathematicians, who believed that there were only two types of space: Euclidean, or flat space, like a table, and spherical space, like a ball. In the nineteenth century, hyperbolic space was discovered – but only in principle. And although mathematicians tried for over a century to find a way to successfully represent this space physically, no one managed it – until Taimina attended that workshop at Cornell. Because as well as being a professor of mathematics, Taimina also liked to crochet.

Taimina learnt to crochet as a schoolgirl. Growing up in Latvia, part of the former Soviet Union, 'you fix your own car, you fix your own faucet – anything', she explains.[4] 'When I was growing up, knitting or any other handiwork meant you could make a dress or a sweater different from everybody else's.' But while she had always seen patterns and algorithms in knitting and crochet, Taimina had never connected this traditional, domestic, feminine skill with her professional work in maths. Until that workshop in 1997. When she saw the battered paper approximation Henderson was using to explain hyperbolic space, she realised: I can make this out of crochet.

And so that's what she did. She spent her summer 'crocheting a classroom set of hyperbolic forms' by the swimming pool. 'People walked by, and they asked me, "What are you doing?" And I answered, "Oh, I'm crocheting the hyperbolic plane."'[5] She has now created hundreds of models and explains that in the process of making them 'you get a very concrete sense of the space expanding exponentially. The first rows take no time but the later rows can take literally hours, they have so many stitches. You get a visceral sense of what "hyperbolic" really means.'[6] Just looking at her models did the same for others: in an interview with the *New York Times* Taimina recalled a professor who had taught hyperbolic space for years seeing one and saying, 'Oh, so that's how they look.'[7] Now her creations are the standard model for explaining hyperbolic space.

Taimina's fundamental contribution to the study of the hyperbolic plane does not, of course, close a data gap that directly relates to women. What this story shows instead is that the case for closing the gender data gap extends beyond women's rights. Closing the data gap, as we've seen from the impact women have in politics, in peace talks, in design and urban planning, is good for everyone. Even mathematicians.

When we exclude half of humanity from the production of knowledge we lose out on potentially transformative insights. Would male mathematicians have come up with Taimina's elegantly simple solution on their own? Unlikely, given how few men are keen crocheters. But in Taimina the traditionally feminine skill of crochet collided with the traditionally masculine sphere of maths. And it was this collision that led to the problem that many mathematicians had given up on as a lost cause finally being solved. Taimina provided the link the male mathematicians were missing.

All too often, however, we don't allow women to provide that link. And so we continue to treat too many of the world's problems as insoluble. Like Freud, we continue to 'knock our heads' against what seem like riddles. But what if, like representing the hyperbolic plane, these problems aren't insoluble? What if, like the problems in broadcast science competitions, all they are missing is a female perspective? The data that we do have is unarguable: as we continue to build, plan and develop our world, we have to start taking account of women's lives. In particular, we have to start accounting for the three themes that define women's relationship with that world.

The first of these themes is the female body – or, to be precise – its invisibility. Routinely forgetting to accommodate the female body in design – whether medical, technological or architectural – has led to a world that is less hospitable and more dangerous for women to navigate. It leads to us injuring ourselves in jobs and cars that weren't designed for our bodies. It leads to us dying from drugs that don't work. It has led to the creation of a world where women just don't *fit* very well.

There is an irony in how the female body is apparently invisible when it comes to collecting data, because when it comes to the second trend that defines women's lives, the visibility of the female body is key. That trend is male sexual violence against women – how we don't measure it, don't design our world to account for it, and in so doing, allow it to limit women's liberty. Female biology is not the reason women are raped. It is not the reason women are intimidated and violated as they navigate public spaces. This happens not because of sex, but because of gender: the social meanings we have imposed on male and female bodies. In order for gender to work, it must be obvious which bodies elicit which treatment. And, clearly, it is: as we've seen, 'the mere sight of a woman' is enough for the viewer to 'immediately elicit a specific set of associated traits and attributions'.[8] To immediately class her as someone to speak over. Someone to cat call. Someone to follow. Someone to rape.

Or maybe just someone to make the tea. Which is where we run into the third trend, which is perhaps the most significant in terms of its impact on women's lives worldwide: unpaid care work. Women are doing far and away more than our fair share of this work – this necessary work without which our lives would all fall apart. And, as with male violence against women, female biology is not the reason women are the bum-wiping class. But recognising a child as female is the reason she will be brought up to expect and accept that as her role. Recognising a woman as female is the reason she will be seen as the appropriate person to clear up after everyone in the office. To write the Christmas and birthday cards to her husband's family – and look after them when they get sick. To be paid less. To go part-time when they have kids.

Failing to collect data on women and their lives means that we continue to naturalise sex and gender discrimination – while at the same time somehow not seeing any of this discrimination. Or really, we don't see it *because* we naturalise it – it is too obvious, too commonplace, too much just the way things are to bother commenting on. It's the irony of being a woman: at once hyper-visible when it comes to being treated as the subservient sex class, and invisible when it counts – when it comes to being counted.

There is one more trend I kept coming across while writing this book: the excuses. Chief amongst these is that women are just too complicated to measure. Everyone was saying this, from transport planners, to medical researchers, to tech developers: they were all knocking their heads up against Freud's riddle of femininity and coming away baffled and defeated. Female bodies are too unharmonious, too menstrual and too hormonal. Women's travel patterns are too messy, their work schedules are too aberrant, their voices are too high. Even when, in the early twentieth century, influential Swiss architect Le Corbusier was devising a standard human model for use in architecture, the female body was 'only belatedly considered and rejected as a source of proportional harmony',[9] with humanity instead represented by a six-foot man with his arm raised (to reach that top shelf I can never reach).

The consensus is clear: women are abnormal, atypical, just plain wrong. Why can't a woman be more like a man? Well, apologies on behalf of the female sex for being so mysterious, but no, we aren't and no we can't. And that is a reality that scientists, politicians and tech bros just need to face up to. Yes, simple is easier. Simple is cheaper. But simple doesn't reflect reality.

Back in 2008, Chris Anderson, then editor of tech magazine *Wired*, penned an article headlined 'The End of Theory: The Data Deluge Makes the Scientific Model Obsolete'.[10] We can 'stop looking for models', Anderson claimed. There is now a better way. Petabytes [that's 1,000 million million bytes to you and me] allow us to say: 'Correlation is enough.' We didn't need to hypothesise about anything, we just needed to crunch the numbers – or, more accurately, 'let statistical algorithms' crunch the numbers. In the era of Trump, Brexit and Cambridge Analytica, this seems Pollyanna-ish to say the least, but even before these data scandals it should have been obvious that his claims were hubristic, because back in 2008 we had even less data on women than we have now. And when you're missing out half the global population in the numbers you feed your statistical algorithms, what you're actually creating is just a big mess.

Anderson holds up Google as an exemplar of what he dubbed 'The Petabyte Age', singing the praises of its 'founding philosophy' that 'we don't know why this page is better than that one: If the statistics of incoming links say it is, that's good enough. No semantic or causal analysis is required. That's why Google can translate languages without actually knowing them (given equal corpus data, Google can translate Klingon into Farsi as easily as it can translate French into German).' Except, as we've seen, Google actually can't translate very well at all, even ten years later. That is, if you care about women being erased from language.

So. Not so simple after all.

Anderson is right about one thing though. There is a better way. And it's a pretty simple one: we must increase female representation in all spheres of life. Because as more women move into positions of power or influence, there's another pattern that is becoming even more apparent: women simply don't forget that women exist as easily as men often seem to.

Women in the film industry are more likely to employ women.[11] Female journalists are significantly more likely to centre a female perspective and to quote women.[12] Female authors do the same: 69% of US female biographers wrote about female subjects in 2015, compared to 6% of male biographers.[13] The emphasis by women on female voices and perspectives extends to the academy. Between 1980 and 2007, female history faculty in the US rose from 15% to 35%[14] – meanwhile across a similar time period (1975–2015), US history faculty specialising in women's history rose from 1% to 10%[15] – a tenfold increase. Female academics are also more likely to assign female authors to their students.[16]

Then there's how women might interpret history: in a 2004 *Guardian* article comedian Sandi Toksvig wrote about how when she was studying anthropology at university one of her female professors held up a photograph of an antler bone with twenty-eight markings on it. 'This,' she said, 'is alleged to be man's first attempt at a calendar.' We all looked at the bone in admiration. 'Tell me,' she continued, 'what man needs to know when 28 days have passed? I suspect that this is woman's first attempt at a calendar.'[17]

When Britain's EU Withdrawal Bill was announced in 2017, the Human Rights Act was explicitly excluded from alteration – but it took a woman, Maria Miller, the Conservative MP for Basingstoke, to force the government to agree to make a statement requiring that Brexit is also compatible with the Equalities Act.[18] Without this concession, a whole range of women's rights could be scrapped after Brexit, with no avenue for legal redress. In the workplace it is often women, like developmental biologist Christiane Nüsslein-Volhard with her foundation to help female PhD students with children, who are putting in place solutions to structural male bias – a bias which male leaders have overlooked and ignored for decades.

Women are also leading the way when it comes to closing the gender data gap. A recent analysis of 1.5 million papers published between 2008–15 found that the likelihood of a study involving gender and sex analysis 'increases with the proportion of women among its authors'[19]. The effect is particularly pronounced if a woman serves as a leader of the author group. This concern for women's health also extends to the political sphere: it took a woman (Paula Sherriff, the Labour MP for Dewsbury) to set up the UK's first All-Party Parliamentary group for women's health in 2016. It was two rogue female Republicans who scotched Donald Trump's attempts to repeal Obamacare (which would have disproportionately impacted on women), voting three times against his proposals.[20]

And women are making a difference in politics more generally. It was two women, Melinda Gates and Hillary Clinton, who spearheaded the UN-backed organisation Data2x that is aimed specifically at closing the global gender data gap. It was a woman, Hillary Clinton, who insisted on going to Beijing in 1995 to make the now famous declaration that 'Human rights are women's rights, and women's rights are human rights.'

And when the worst happens, women are there too, filling in the gaps left by male-biased disaster relief. Researchers found that the 'masculine and muscular image[s] of relief workers' that dominated the media post-Katrina were belied by women who were 'working tirelessly and courageously' behind the scenes.[21] The same thing has happened in Puerto Rico, all but abandoned by the US government after Hurricane Maria devastated the region in 2017. 'The reality is that when you go to communities, mostly it is women as leaders and as community organizers,' Adi Martínez-Román, executive director for a non-profit that provides legal assistance to low-income families, told journalist Justine Calma.[22] These women have collected data by 'wad[ing] into flooded neighbourhoods' and canvassing the abandoned communities.[23] And they have developed and provided evidence-based solutions. They've set up soup kitchens. They've raised money and rebuilt roads. They've distributed 'solar-powered lights, generators, gas, clothes, shoes, tampons, batteries, medication, mattresses, water'. They set up 'free legal aid societies to help families navigate the confusing and ill-designed processes required to file FEMA claims'. They've even managed to source some communal, solar-powered washing machines.

The solution to the sex and gender data gap is clear: we have to close the female representation gap. When women are involved in decision-making, in research, in knowledge production, women do not get forgotten. Female lives and perspectives are brought out of the shadows. This is to the benefit of women everywhere, and as the story of Taimina, the crocheting maths professor shows, it is often to the benefit of humanity as a whole. And so, to return to Freud's 'riddle of femininity', it turns out that the answer was staring us in the face all along. All 'people' needed to do was to ask women.

Epilogue

The COVID-19 outbreak officially became a pandemic almost exactly a year after *Invisible Women* first came out. As the world went into lockdown, I watched with horror as the message of the book became acutely relevant in a way I had never seriously imagined. In the preceeding pages you have read about a widespread societal bias that frames men as the default, neutral humans, and how this default male bias has led to data gaps that, at their most serious, can prove fatal for women. You also read about how in times of crisis, these gaps and biases are exacerbated – all in the service of focusing on "the real problem."

Well, since March 2020, we have been living through a time of global crisis. And there are plenty of problems to go around.

Let's start with our continual failure to systematically collect and publish sex-disaggregated data on symptoms, infection rates, and death rates from COVID-19. This was a disease which, from very early on, made its sex-specific effects clear. Men were dying at about a 2:1 rate to women. And we didn't know why, because we weren't collecting the data. Who was getting tested? Who was going on to test positive? We needed this data to make sense of the death rates. Without it, we couldn't say if men were more likely to *get* COVID, or more likely to *die* from COVID. And without knowing *that*, we didn't know where to direct resources to save the most lives.

By the end of March only six out of the twenty most affected countries[1] were publishing properly sex-disaggregated data. It took until April for the US and the UK to start reporting sex-disaggregated data on deaths, and May for them to report sex-disaggregated data on cases.

By September 2020, only 30% of the world's countries are reporting sex-disaggregated data on cases and deaths. Fewer than half of high-income countries are publishing this data. One in three countries are not reporting any sex-disaggregated health data at all, and some countries that *were* reporting sex-disaggregated data have since stopped.[2] And around the world, including here in London, where I'm writing this, the R number – the reproduction number – is rising.

Meanwhile, our historic failure to study the female body, in preference for the "simpler" default-male body, is hampering our hunt for a cure. Once we had isolated the female immune system as a significant factor in women's lower death rate, we quickly ran into the barrier of how little we know about that immune system. What made it so much better at fighting off this virus? Was it sex hormones? Was it genes? Was it both? We simply don't know. We haven't, on the whole, done the research.

Still, thanks to the work of some pioneering researchers, we do now know a little. In *Invisible Women* I reported on a 2016 cell study which found that while female cells exposed to oestrogen were better able to fight off a virus, the same did not hold for male cells (see p. 207). Nevertheless, in our desperation to find a cure, a hospital on Long Island, NY, started injecting their male COVID patients with oestrogen to see if that might save their lives.[3] The outcome of that experiment is still pending at the time of writing.

Meanwhile, COVID-19 research (including on immunity[4]) continues to be published that does not sex-disaggregate its data, or even specify the sex of its participants.[5] Which, given the by now well-known sex differences in immune function is, to say the least, odd.

This failure to sex-disaggregate, or to adequately account for women, isn't only visible in studies specific to the disease itself. A recent study tested mask fit on a single male participant – described in the paper as "a common individual" – with a female participant used for comparison on a few masks. This is a particularly insidious example of the "default male" or "reference man" tendency you have read about in the following pages, because, in a healthcare setting, if any sex is the default it is women. In the US, women hold 76% of healthcare jobs.[6] This figure rises to 85%[7] when it comes to nurses, and 89%[8] for home care workers.

And it is women who have been struggling for years with Personal Protective Equipment (PPE) that does not fit their bodies, because it was designed for the default male. I wrote about this problem in chapter 5 so when the messages about ill-fitting PPE started pouring in from women on the frontlines, I was not surprised. What did surprise me was the reluctance of manufacturers, politicians, or even journalists, to acknowledge or address this issue, despite the fact that healthcare practitioners are some of the most at-risk workers in the world right now.

Naturally, we aren't collecting data on the extent of the problem: I sent Freedom of Information requests to 300 of the UK's National Health Service providers to ask them how many female versus male healthcare workers were failing their fit tests (the test used to determine if a mask fits well enough to protect you). The vast majority were not collecting sex-disaggregated data. For those that were, women dominated failure rates – no surprise given 77% of UK healthcare workers are female.

Meanwhile, I was being contacted regularly by healthcare workers who consistently told the same story: women were failing mask fit tests because their faces were "too small" – or as I like to call them, perfectly average size for a woman. Despite this, and despite a mass email-writing campaign to MPs, there has still been no serious recognition that this is a problem.

Outside of healthcare, the social and economic impact of the COVID-19 pandemic is hitting women hard. Women – especially women of colour – dominate in high-risk low-paid jobs, and around the world they also dominate among those who have had to reduce their hours, who have been furloughed, or who have lost their jobs altogether.[9] This is partly because of the sectors women are more likely to work in, but it is also because, as you have discovered in these pages, women still shoulder 75% of the world's unpaid care work. And with children out of nursery and school, and little to no help in the home, that 75% share has suddenly got an awful lot weightier. And yes, many men have stepped up, but the stats don't lie:[10] this has been a disproportionately female issue.

The result is that women's productivity has taken a massive hit[11] – although even here, we aren't always collecting sex-disaggregated data, resulting in research that claims gender-neutral productivity is up.[12] Meanwhile, the UK government quietly dropped the requirement for companies to publish data on their gender pay gap this year,[13] and countries around the world are telling workers to return to the office without investing in the childcare that will enable them to do that.[14] But without investment in social care infrastructure (see chapter 12) the much-sought after economic recovery simply won't happen,[15] no matter how much we "Build, Build, Build".[16]

If we carry on down our current path, things do not look rosy for women – or for anyone. A successful economy without female paid labour is simply not viable in the 21st century. But there is another future. A future in which we learn from the lessons of this pandemic: that care work is indispensable to our economy; and that sex-disaggregated data is indispensable to our health. In that future, we invest in the (currently unpaid or low-paid) care work that has been invisibly propping up our formal economies for centuries; and we learn from the catastrophic misconception that collecting data on men is enough. In that future we not only recover from this pandemic, we emerge from it stronger and more resilient than ever. I know which path I would choose.

Acknowledgements

Writing a book can feel like a lonely endeavour and often it is. But it's also in many ways a group achievement. My first thanks have to go to Rachel Hewitt, who introduced me to her, now my, amazing agent Tracy Bohan at the Wylie Agency, because without that introduction this book would probably never have happened. And Tracy has been a dream to work with. I'm so grateful to her for taking me on and helping me to shape a book proposal that got me my very first book auction – not to mention always being on hand to very calmly, politely and Canadianly, deal with every problem (including those of my own making) that I've thrown at her. Thanks too to her wonderful assistant Jennifer Bernstein who has been so supportive throughout.

Next thanks go to my two brilliant editors, Poppy Hampson and Jamison Stoltz, both of whom immediately got the idea in a way no-one else did. They have been painstaking and methodical, taking me carefully through the various drafts, asking questions that forced me to sharpen my argument and defend my thesis. This book is what it is because of them, and I'm so grateful to them for challenging me to make it better. Special thanks to Poppy for having at least two crisis coffees with me as I had minor breakdowns about Never Finishing. And huge thanks also to all at Chatto & Windus and at Abrams Books for taking this on and being so dedicated to making it work from the very beginning.

I have so many people to thank who were generous with their time and expertise. Nishat Siddiqi for giving me a crash-course in how the heart works and answering all my no doubt ridiculous questions about the cardiovascular system. James Ball who did the same with all my stats questions alongside being a brilliant friend who listened to my more or less daily wails about getting to the end. Thanks too to my lovely friend Alex Kealy who was my other go-to for stats and also had to put up with semi-regular wailing. Alex Scott gets special mention for being amazingly kind and reading through my medical chapters to make sure I hadn't made any howlers, as does Greg Callus who did a legal fact-check for me. I'd also like to thank Ellie Cosgrave for talking to me about gender and engineering and in particular for introducing me to the gender and disasters crowd, without which this book would be incomplete.

Special acknowledgement has to go to Helen Lewis for her spot on coinage 'vomit draft' which I found incredibly useful to hold in mind as a way to just get the initial words down. Huge thanks also to her, Sarah Ditum, Alice Ford, Nicky Woolf and Luke McGee for bravely reading some very early sections (and particularly to Helen for turning her expert eye to some particularly knotty sections). I hope none of you emerged too traumatised from the experience.

To all my lovely friends for supporting me and putting up with my disappearing for months on end and repeatedly cancelling plans: thank you for your patience and support and thank you for listening. I couldn't ask for a better bunch and I'm so grateful to have all of you in my life, especially my beloved HarpySquad and the gang of 🦆 who really have had to suffer with me through this book on a daily basis. You know who you are.

Biggest thanks of all, though, have to go to my amazing Official Friend and cheerleader Tracy King, who has not only worked with me on my madcap feminist campaigns, but who read the very earliest vomit drafts of this book and never stopped encouraging me and promising me I would eventually finish. I could never have done this and have remained (relatively) sane without her.

OK, there is one more thanks: to my beloved dog Poppy. She really does make the work that I do possible – not just by sitting on my lap, but also by distracting me when I've been typing for too long. She literally just licked my arm as I typed that. She's the gorgeous best and I couldn't do anything without her.

Endnotes

Preface

1 Beauvoir, Simone de (1949), *The Second Sex*, Parshley, H.M. trans. (1953), London

Introduction

1 http://science.sciencemag.org/content/164/3883/1045.1
2 Slocum, Sally (1975), 'Woman the gatherer: male bias in anthropology', in Reiter, Rayna R. ed. (1975), *Toward an Anthropology of Women*. Monthly Review Press
3 http://www.independent.co.uk/news/science/human-evolution-violence-instinct-to-kill-murder-each-other-a7335491.html
4 https://www.nature.com/nature/journal/v538/n7624/full/nature19758.html
5 https://www.eurekalert.org/pub_releases/2016–06/uog-mdb061716.php
6 http://www.smh.com.au/lifestyle/news-and-views/social/no-women-arent-as-likely-to-commit-violence-as-men-20141118-3km9x.html
7 https://www.ons.gov.uk/peoplepopulationandcommunity/crimeandjustice/compendium/focusonviolentcrimeandsexualoffences/yearendingmarch2015/chapter2homicide#focus-on-domestic-homicides
8 https://www.bjs.gov/content/pub/pdf/htus8008.pdf

9 http://www.unodc.org/documents/gsh/pdfs/2014_GLOBAL_HOMICIDE_BOOK_web.pdf

10 https://news.nationalgeographic.com/news/2013/10/131008-women-handprints-oldest-neolithic-cave-art/

11 https://www.theguardian.com/science/2017/sep/15/how-the-female-viking-warrior-was-written-out-of-history

12 https://news.nationalgeographic.com/2017/09/viking-warrior-woman-archaeology-spd/

13 https://news.nationalgeographic.com/2017/09/viking-warrior-woman-archaeology-spd/

14 https://www.nytimes.com/2017/09/14/world/europe/sweden-viking-women-warriors-dna.html

15 https://www.nytimes.com/2017/09/14/world/europe/sweden-viking-women-warriors-dna.html

16 Walker, Phillip (1995), 'Problems of Preservation and Sexism in Sexing: Some Lessons from Historical Collections for Palaeodemographers', in Saunders, S. R. and Herring A. (eds.), *Grave Reflections, Portraying the Past through Cemetery Studies* (Canadian Scholars' Press, Toronto); https://namuhyou.wordpress.com/2016/06/18/sexism-when-sexing-your-skull-cultural-bias-when-sexing-the-skull/

17 https://www.nytimes.com/2017/09/14/world/europe/sweden-viking-women-warriors-dna.html

18 https://www.theguardian.com/commentisfree/2017/sep/18/battle-prejudice-warrior-women-ancient-amazons

19 https://www.foreignaffairs.com/articles/2015–05-06/warrior-women

20 Hegarty, Peter and Buechel, Carmen (2006), 'Androcentric Reporting of Gender Differences', *APA Journals: 1965–2004 Review of General Psychology*, 10:4, 377–89;
Vainapel, Sigal, Shamir, Opher Y., Tenenbaum, Yulie and Gilam, Gadi (2015), 'The Dark Side of Gendered Language: The Masculine-Generic Form as a Cause for Self-Report Bias', *Psychological Assessment Issue*, 27:4, 1513–19;
Sczesny, Sabine, Formanowicz, Magda, and Moser, Franziska (2016), 'Can Gender-Fair Language Reduce Gender Stereotyping and Discrimination?', *Frontiers in Psychology*, 7, 1–11;
Horvath, Lisa Kristina and Sczesny, Sabine (2016), 'Reducing women's lack of fit with leadership positions? Effects of the wording of job advertisements', *European Journal of Work and Organizational Psychology*, 25:2, 316–28;
Stout, Jane G. and Dasgupta, Nilanjana (2011), 'When He Doesn't Mean

You: Gender-Exclusive Language as Ostracism', *Personality and Social Psychology Bulletin*, 36:6, 757–69;

Vervecken, Dries, Hannover, Bettina and Wolter, Ilka (2013), 'Changing (S) expectations: How gender fair job descriptions impact children's perceptions and interest regarding traditionally male occupations', *Journal of Vocational Behavior*, 82:3, 208–20;

Prewitt-Freilino, J. L., Caswell, T. A. and Laakso, E. K. (2012), 'The Gendering of Language: A Comparison of Gender Equality in Countries with Gendered, Natural Gender, and Genderless Languages', *Sex Roles*, 66: 3–4, 268–81;

Gygax, Pascal, Gabriel, Ute, Sarrasin, Oriane, Oakhill, Jane and Garnham, Alan (2008), 'Generically intended, but specifically interpreted: When beauticians, musicians, and mechanics are all men', *Language and Cognitive Processes*, 23:3, 464–85;

Stahlberg, D., Sczesny, S. and Braun, F. (2001), 'Name your favorite musician: effects of masculine generics and of their alternatives in German', *Journal of Language and Social Psychology*, 20, 464–69

21 Stahlberg, Sczesny and Braun (2001)

22 Sczesny, Formanowicz and Moser (2016); Vervecken, Hannover and Wolter (2013)

23 Stahlberg, D. and Sczesny, S. (2001), 'Effekte des generischen Maskulinums und alternativer Sprachformen auf den gedanklichen Einbezug von Frauen' [The impact of masculine generics on the cognitive inclusion of women], *Psychol. Rundsch.*, 52, 131–40; Horvath and Sczesny (2016); Sczesny, Formanowicz and Moser (2016)

24 Stout and Dasgupta (2011); Sczesny, Formanowicz and Moser (2016)

25 Gygax, Gabriel, Sarrasin, Oakhill and Garnham (2008)

26 Vainapel, Shamir, Tenenbaum and Gilam (2015)

27 Ignacio Bosque, 'Sexismo lingüístico y visibilidad de la mujer', http://www.rae.es/sites/default/files/Sexismo_linguistico_y_visibilidad_de_la_mujer_0.pdf

28 Vainapel, Shamir, Tenenbaum and Gilam (2015)

29 https://www.theguardian.com/uk-news/2018/feb/01/dany-cotton-london-fire-chief-sexist-abuse-over-firefighter-sam-campaign

30 Horvath and Sczesny (2016)

31 Ibid.

32 Ibid.

33 Prewitt-Freilino, Caswell and Laakso (2012)

34 https://www.emogi.com/insights/view/report/1145/2016-emoji-report

35 http://www.adweek.com/digital/report-92-of-online-consumers-use-emoji-infographic/

36 https://unicode.org/L2/L2016/16160-emoji-professions.pdf

37 http://www.adweek.com/digital/report-92-of-online-consumers-use-emoji-infographic/

38 http://www.unicode.org/L2/L2016/16181-gender-zwj-sequences.pdf

39 Bradley, Adam, MacArthur, Cayley, Carpendale, Sheelagh and Hancock, Mark, 'Gendered or Neutral? Considering the Language of HCI', Graphics Interface Conference 2015, 3–5 June, Halifax, Nova Scotia, Canada, http://graphicsinterface.org/wp-content/uploads/gi2015-21.pdf

40 https://genderedinnovations.stanford.edu/institutions/bias.html

41 Naureen Durrani (2008), 'Schooling the 'other': the representation of gender and national identities in Pakistani curriculum texts', *Compare: A Journal of Comparative and International Education*, 38:5, 595–610

42 Lambdin, Jennifer R., Greer, Kristen M., Jibotian, Kari Selby, Wood, Kelly Rice and Hamilton, Mykol C. (2003), 'The Animal = Male Hypothesis: Children's and Adults' Beliefs About the Sex of Non-Sex-Specific Stuffed Animals', *Sex Roles*, 48:11–12, 471–482

43 http://www.br-online.de/jugend/izi/deutsch/forschung/gender/IZI_Guidelines_WEB.pdf

44 http://seejane.org/wp-content/uploads/key-findings-gender-roles-2013.pdf

45 http://wmc.3cdn.net/dcdb0bcb4b0283f501_mlbres23x.pdf

46 http://www.news.com.au/finance/money/australia-a-world-leader-in-female-representation-on-banknotes/news-story/3cf7c3b5ed3838075d571a64c7fcdff6

47 http://cdn.agilitycms.com/who-makes-the-news/Imported/reports_2015/highlights/highlights_en.pdf

48 Silvina Bongiovanni (2014), "No se preocupe la señora marquesa': A study of gender bias in example sentences in the RAE grammar textbook', *IULC Working Papers*, 14:1 https://www.indiana.edu/~iulcwp/wp/article/viewFile/14-05/146

49 Clark, Roger, Allard, Jeffrey and Mahoney, Timothy (2004) 'How Much of the Sky? Women in American High School History Textbooks from the 1960s, 1980s and 1990s', *Social Education*, 68:1, 57–62

50 Amy L. Atchison (2017), 'Where Are the Women? An Analysis of Gender Mainstreaming in Introductory Political Science Textbooks', *Journal of Political Science Education*, 13:2, 185–199

51 Iveta Silova (2016), 'Gender Analysis of Armenian School Curriculum and Textbooks Policy Brief', PhD (June 2016), Arizona State University, https://openknowledge.worldbank.org/bitstream/handle/10986/24948/Gender0analysi0ooks000policy0brief.pdf?sequence=1&isAllowed=y; Chiponda, Annie F and Wassermann, Johann (2016), 'The depiction of women in the verbal text of a junior secondary Malawian history textbook – an analysis', *Yesterday & Today*, 16, 40–59; https://ei-ie.org/en/woe_homepage/woe_detail/15405/curriculum-textbooks-and-gender-stereotypes-the-case-of-pakistan; Durrani (2008); Ullah, Hazir and Skelton, Christine (2013), 'Gender representation in the public sector schools textbooks of Pakistan', *Educational Studies* 39:2; 2006, 2007, 2009 and 2010 studies cit. Chiponda, Annie F and Wassermann, Johann (2016)

52 http://www.siliconera.com/2016/12/02/metroid-developers-discuss-decided-make-samus-aran-woman-new-interview/

53 http://www.pewinternet.org/2015/12/15/gaming-and-gamers/

54 http://wmc.3cdn.net/dcdb0bcb4b0283f501_mlbres23x.pdf

55 https://feministfrequency.com/2015/06/22/gender-breakdown-of-games-showcased-at-e3-2015/

56 http://www.kotaku.co.uk/2015/07/15/fifas-struggle-to-include-women-reveals-a-lot-about-gamings-problems-with-diversity

57 https://feministfrequency.com/2016/06/17/gender-breakdown-of-games-showcased-at-e3-2016/

58 http://www.nytimes.com/1991/03/03/nyregion/campus-life-georgetown-white-male-writers-is-the-title-of-english-112.html

59 https://www.theguardian.com/film/2015/sep/05/suffragette-review-historical-drama-tub-thumps-hard-despite-having-your-vote

60 https://ai2-s2-pdfs.s3.amazonaws.com/05e1/0638aab94ca0d46ddde8083ff69859a0401e.pdf

61 https://www.theguardian.com/lifeandstyle/womens-blog/2016/aug/17/normal-society-means-male-andy-murray-venus-serena-williams?CMP=fb_gu

62 https://www.theguardian.com/football/2015/jul/05/usa-womens-world-cup-champions-japan

63 http://www.independent.co.uk/arts-entertainment/books/news/thor-as-woman-marvel-reveals-new-incarnation-of-superhero-in-comic-series-9608661.html

64 https://www.wired.com/2015/10/hugo-awards-controversy/

65 http://www.mamamia.com.au/star-wars-movie-features-a-female-lead.,; http://screencrush.com/rogue-one-female-lead-angry-fans/

66 http://www.telegraph.co.uk/news/2017/07/21/former-doctor-peter-davison-says-female-choice-role-means-loss/
67 http://uk.businessinsider.com/doctor-who-first-woman-jodie-whittaker-sexist-reactions-2017–7
68 https://www.theguardian.com/books/2014/nov/25/readers-prefer-authors-own-sex-goodreads-survey
69 https://kotaku.com/ubisoft-cut-plans-for-female-assassins-in-unity-1589278349
70 http://www.kotaku.co.uk/2014/06/16/whole-assassins-creed-thing
71 For more, see Anna Beer (2016): *Sounds and Sweet Airs: The Forgotten Women of Classical Music*, London
72 http://www.bbc.co.uk/news/entertainment-arts-39191514
73 https://www.theguardian.com/commentisfree/2017/sep/18/battle-prejudice-warrior-women-ancient-amazons
74 https://www.theguardian.com/world/2017/feb/01/caroline-louisa-daly-art-men-attribution
75 https://news.nationalgeographic.com/news/2013/13/130519-women-scientists-overlooked-dna-history-science/
76 http://www.newn.cam.ac.uk/about/history/biographies/
77 Beer (2016).
78 Despite being lauded as a child prodigy, Fanny Hensel was informed by her father that 'Music will perhaps become his [her brother, Felix Mendelssohn] career, whilst for you it can and must only be an ornament.'
79 http://www.telegraph.co.uk/women/womens-life/9790633/Will-Goves-posh-white-blokes-history-curriculum-ignore-women.html
80 www.telegraph.co.uk/education/educationopinion/9973999/Sorry-NUT-Goves-history-reforms-are-no-pub-quiz.html
81 http://www.telegraph.co.uk/culture/tvandradio/5077505/History-has-been-feminised-says-David-Starkey-as-he-launches-Henry-VIII-series.html
82 https://teachingwomenshistory.com/teaching-resources/medieval-women/
83 https://www.nytimes.com/2016/11/20/opinion/sunday/the-end-of-identity-liberalism.html?_r=0
84 http://www.wbur.org/politicker/2016/11/21/bernie-sanders-berklee
85 http://thehill.com/homenews/campaign/307014-sanders-dems-must-move-beyond-identity-politics
86 http://www.theaustralian.com.au/opinion/columnists/paul-kelly/donald-trumps-election-a-rejection-of-identity-politics/news-story/147b11c08b64702d3f9be1821416cb72

87 https://twitter.com/RichardBurgon/status/822417591713075201
88 https://www.theguardian.com/commentisfree/2016/dec/01/blame-trump-brexit-identity-liberalism
89 https://www.bls.gov/oes/current/naics4_212100.htm#00–0000
90 https://www.bls.gov/oes/current/oes372012.htm
91 Bourdieu, Pierre (1972) *Outline of a Theory of Practice*, Nice, Richard trans. (1977), Cambridge
92 http://theconversation.com/donald-trump-and-the-rise-of-white-identity-in-politics-67037
93 http://www.vox.com/2016/11/1/13480416/trump-supporters-sexism

Chapter 1

1 https://openknowledge.worldbank.org/bitstream/handle/10986/28542/120500.pdf?sequence=6
2 http://planphilly.com/articles/2015/01/26/septa-has-largest-percentage-of-female-riders-64-among-large-transit-agencies?utm_content=buffer97258&utm_medium=social&utm_source=twitter.com&utm_campaign=buffer
3 Ceccato, Vania (2017), 'Women's victimisation and safety in transit environments', *Crime Prevention and Community Safety*, 19:3–4, 163–7
4 http://ec.europa.eu/commfrontoffice/publicopinion/archives/ebs/ebs_422a_en.pdf; World Bank (2007), *Gender and Urban Transport: Fashionable and Affordable Module 7a Sustainable Transport: A Sourcebook for Policy-makers in Developing Cities*
5 http://www.wnyc.org/story/283137-census-data-show-public-transit-gender-gap/
6 Ceccato (2017)
7 http://content.tfl.gov.uk/travel-in-london-understanding-our-diverse-communities.pdf 2015
8 http://content.tfl.gov.uk/gender-equality-scheme-2007–2010.pdf
9 Sánchez de Madariaga, Inés, 'Mobility of Care: Introducing New Concepts in Urban Transport', in Roberts, Marion and Sánchez de Madariaga, Inés (eds.) (2013), *Fair Shared Cities: The Impact of Gender Planning in Europe*, Farnham
10 http://media.leidenuniv.nl/legacy/leru-paper-gendered-research-and-innovation.pdf
11 http://ssmon.chb.kth.se/volumes/vol16/5_Rolfsman_Bylund.pdf 2012

12 https://lucris.lub.lu.se/ws/files/6151586/2295991.pdf
13 http://media.leidenuniv.nl/legacy/leru-paper-gendered-research-and-innovation.pdf
14 www.chicksontheright.com/feminist-snow-plowing-disrupts-traffic-and-normal-life-for-people-in-sweden/; https://heatst.com/world/feminist-snow-plowing-system-brings-stockholm-to-a-standstill/
15 https://heatst.com/world/feminist-snow-plowing-system-brings-stockholm-to-a-standstill/
16 http://www.dn.se/arkiv/stockholm/jamstalld-snorojning-blev-ett-fiasko-i-ovadret/
17 http://thecityfix.com/blog/brasilia-brazil-women-bus-stop-night-safety-sexual-assault-luisa-zottis/
18 http://ec.europa.eu/commfrontoffice/publicopinion/index.cfm/ResultDoc/download/DocumentKy/61244
19 Sánchez de Madariaga (2013)
20 https://www.newstatesman.com/politics/uk/2017/07/unless-living-standards-improve-theresa-mays-cameron-tribute-act-will-continue
21 https://www.unison.org.uk/content/uploads/2014/06/On-line-Catalogue224222.pdf
22 https://www.itdp.org/wp-content/uploads/2014/07/7aGenderUTSept300.pdf; World Bank (2007)
23 Review of World Bank infrastructure projects 1995–2009 http://siteresources.worldbank.org/EXTSOCIALDEVELOPMENT/Resources/244362-1265299949041/6766328-1270752196897/Gender_Infrastructure2.pdf
24 Sánchez de Madariaga (2013); Tran, Hoai Anh and Schlyter, Ann (2010), 'Gender and class in urban transport: the cases of Xian and Hanoi', *Environment and Urbanization*, 22:1, 139–55
25 http://wricitieshub.org/sites/default/files/Final_Report_24082015_0.pdf; http://content.tfl.gov.uk/travel-in-london-understanding-our-diverse-communities.pdf
26 http://content.tfl.gov.uk/travel-in-london-understanding-our-diverse-communities.pdf
27 http://genderedinnovations.stanford.edu/case-studies/urban.html
28 https://tfl.gov.uk/campaign/hopper-fare
29 http://humantransit.org/2010/02/the-power-and-pleasure-of-grids.html
30 http://humantransit.org/2014/08/charging-for-connections-is-insane.html
31 https://las.depaul.edu/centers-and-institutes/chaddick-institute-for-metropolitan-development/research-and-publications/Documents/Have%20App%20Will%20Travel%20Uber%20-%20CTA.pdf

32 Ibid.
33 http://webfoundation.org/docs/2015/10/womens-rights-online_Report.pdf
34 http://www3.weforum.org/docs/GGGR16/WEF_Global_Gender_Gap_Report_2016.pdf
35 http://conversableeconomist.blogspot.co.uk/2015/10/unpaid-care-work-women-and-gdp.html
36 World Bank (2007)
37 https://www.gov.uk/government/uploads/system/uploads/attachment_data/file/576095/tsgb-2016-report-summaries.pdf
38 http://wricitieshub.org/sites/default/files/Final_Report_24082015_0.pdf 2015 Bhopal
39 http://civitas.eu/sites/default/files/civ_pol-an2_m_web.pdf
40 https://www.rita.dot.gov/bts/sites/rita.dot.gov.bts/files/TSAR_2016r.pdf
41 Sánchez de Madariaga (2013)
42 http://hdr.undp.org/sites/default/files/chapter4.pdf
43 http://www.imf.org/external/pubs/ft/sdn/2013/sdn1310.pdf (duffle 2012)
44 http://siteresources.worldbank.org/INTAFRREGTOPGENDER/Resources/gender_econ_growth_ug.pdf
45 https://www.habitatforhumanity.org.uk/what-we-do/where-we-work/latin-america-and-caribbean/brazil
46 http://abeiradourbanismo.blogspot.co.uk/2012/02/habitacao-emprego-e-mobilidade.html
47 https://lsecities.net/media/objects/articles/relocating-homes-and-lives-in-rios-olympic-city/en-gb/
48 https://www.boell.de/en/2014/06/11/we-were-not-invited-party-women-and-world-cup
49 http://www.rioonwatch.org/?p=6527
50 https://www.lincolninst.edu/sites/default/files/pubfiles/koch_wp13jk1.pdf
51 https://www.boell.de/en/2014/06/11/we-were-not-invited-party-women-and-world-cup
52 https://lsecities.net/media/objects/articles/relocating-homes-and-lives-in-rios-olympic-city/en-gb/
53 http://www.rioonwatch.org/?p=6527
54 http://www.rioonwatch.org/?p=25015
55 https://www.boell.de/en/2014/06/11/we-were-not-invited-party-women-and-world-cup
56 http://www.citylab.com/commute/2013/09/how-design-city-women/6739/
57 Ibid.

58 http://www3.weforum.org/docs/GGGR16/WEF_Global_Gender_Gap_Report_2016.pdf
59 Alexis Grenell (2015), 'Sex & the Stadt: Reimagining Gender in the Built Environment', http://www.academia.edu/10324825/Sex_and_the_Stadt_Reimagining_Gender_in_the_Built_Environment
60 Architekturzentrum Wien (2008), *Housing in Vienna: Innovative, Social, Ecological*, Vienna
61 http://usatoday30.usatoday.com/news/nation/2007-12-25-Designingwomen_N.htm

Chapter 2

1 https://twitter.com/SamiraAhmedUK/status/849338626202886144
2 https://www.barbican.org.uk/about-barbican/people
3 Banks, Taunya Lovell (1991), 'Toilets as a Feminist Issue: A True Story', *Berkeley Women's Law Journal*, 6:2 263–289
4 Greed, Clara (2014), 'Global gendered toilet provision', in 'More Public than Private: Toilet Adoption and Menstrual Hygiene Management II', AAG Annual Conference, Tampa, Florida, USA, 8–12 April 2014
5 https://www.ncbi.nlm.nih.gov/pmc/articles/PMC3749018/
6 Greed (2014)
7 http://www.unric.org/en/latest-un-buzz/29530-one-out-of-three-women-without-a-toilet
8 http://womendeliver.org/2016/yale-study-examines-link-sexual-violence-access-sanitation/
9 http://indianexpress.com/article/india/india-news-india/india-has-60-4-per-cent-people-without-access-to-toilet-study/
10 Greed (2014)
11 Ibid.
12 http://www.huffingtonpost.com/rose-george/open-defecation-india_b_7898834.html https://www.theguardian.com/global-development/2014/aug/28/toilets-india-health-rural-women-safety
13 https://www.hrw.org/sites/default/files/report_pdf/wrdsanitation0417_web_0.pdf 2017
14 Sommer, Marni, Chandraratna, Sahani, Cavill, Sue, Mahon, Therese, and Phillips-Howard, Penelope (2016), 'Managing menstruation in the workplace: an overlooked issue in low- and middle-income countries', *Int. J. Equity Health*, 15:86

15 https://www.hrw.org/sites/default/files/report_pdf/wrdsanitation0417_web_0.pdf 2017
16 http://ohrh.law.ox.ac.uk/bombay-high-court-makes-right-to-clean-toilets-a-fundamental-right-for-women-in-india/
17 https://www.pri.org/stories/2014-11-25/women-india-agitate-their-right-pee
18 Ibid.
19 http://indianexpress.com/article/cities/mumbai/women-in-slums-forced-to-defecate-in-open-say-community-toilets-are-unsafe-at-night/
20 https://www.theguardian.com/global-development/2014/aug/28/toilets-india-health-rural-women-safety; https://womennewsnetwork.net/2012/12/19/india-women-new-delhi-slum-toilets/
21 https://www.newsdeeply.com/womenandgirls/articles/2017/02/03/without-access-clean-safe-toilets-women-face-assault-illness
22 Jadhav, A., Weitzman, A. and Smith-Greenaway, E. (2016), 'Household sanitation facilities and women's risk of non-partner sexual violence in India', *BMC Public Health*, 16:1139
23 https://www.npr.org/sections/parallels/2014/06/02/318259419/double-rape-lynching-in-india-exposes-caste-fault-lines
24 http://www.dnaindia.com/mumbai/report-right-to-pee-bombay-high-court-gives-municipal-corporations-deadline-in-pil-on-toilets-for-women-2045476
25 https://broadly.vice.com/en_us/article/the-women-in-india-fighting-for-the-right-to-pee
26 http://mumbaimirror.indiatimes.com/mumbai/civic/BMCs-promise-for-womens-toilets-goes-down-the-drain/articleshow/50801316.cms
27 http://journals.plos.org/plosone/article?id=10.1371/journal.pone.0122244
28 https://www.pri.org/stories/2014-11-25/women-india-agitate-their-right-pee
29 https://www.newsdeeply.com/womenandgirls/articles/2017/02/03/without-access-clean-safe-toilets-women-face-assault-illness
30 Greed (2014)
31 http://www.phlush.org/wp-content/uploads/2009/02/americanrestroom-calltoactionpaper.pdf
32 https://blogs.ucl.ac.uk/ucloo-festival-2013/2013/09/17/toilets-gender-and-urbanism/
33 http://transweb.sjsu.edu/sites/default/files/2611-women-transportation.pdf 2009
34 http://transweb.sjsu.edu/sites/default/files/2611-women-transportation.pdf

35 Gardner, Natalie, Cui, Jianqiang and Coiacetto, Eddo (2017), 'Harassment on public transport and its impacts on women's travel behaviour', *Australian Planner*, 54:1, 8–15

36 Ibid.

37 Ibid.

38 http://transweb.sjsu.edu/sites/default/files/2611-women-transportation.pdf

39 Gardner, Cui and Coiacetto (2017)

40 Ceccato, Vania and Paz, Yuri (2017), 'Crime in São Paulo's metro system: sexual crimes against women', *Crime Prevention and Community Safety*, 19:3–4, 211–26

41 http://www.cbgaindia.org/wp-content/uploads/2017/01/Women-safety-in-delhi.pdf

42 http://www.hindustantimes.com/delhi-news/need-to-make-public-transport-in-delhi-women-friendly-study/story-Eq8h997zRiq8XTdIr7dQ0H.html

43 Ceccato and Paz (2017)

44 Gardner, Cui and Coiacetto (2017)

45 https://www.bbc.co.uk/news/uk-england-nottinghamshire-44740362

46 Ceccato and Paz (2017)

47 Gardner, Cui and Coiacetto (2017)

48 http://www.huffingtonpost.com/soraya-chemaly/for-women-rape-isnt-a-mom_b_9997350.html

49 http://www.bbc.co.uk/news/uk-england-london-29818435

50 https://www.itdp.org/wp-content/uploads/2017/01/8.-Beyond-the-Women-Only-Train-Car-Gender-and-Sustainable-Transport.pdf

51 Ceccato and Paz (2017)

52 http://www.nytimes.com/2013/05/25/world/americas/rapes-in-brazil-spur-class-and-gender-debate.html?pagewanted=all&_r=0 ; http://thecityfix.com/blog/women-public-safety-demands-yasmin-khan/

53 http://www.unwomen.org/en/news/stories/2016/11/improving-womens-safety-in-mexico-city

54 http://thecityfix.com/blog/women-public-safety-demands-yasmin-khan/

55 https://www.thelocal.fr/20160615/half-of-french-woman-alter-clothes-to-avoid-harassment

56 https://www.thelocal.fr/20160615/half-of-french-woman-alter-clothes-to-avoid-harassment

57 http://www.thehoya.com/metro-surveys-sexual-harassment-cases/

58 http://www.huffingtonpost.com/soraya-chemaly/for-women-rape-isnt-a-mom_b_9997350.html

59 http://www.nbcwashington.com/news/local/Man-Accused-of-Metro-Assault-Was-Indecent-Exposure-Suspect-380782091.html
60 https://www.washingtonpost.com/news/dr-gridlock/wp/2017/10/20/why-the-metoo-movement-is-a-public-transportation-issue/?utm_term=.09b8335a38b6
61 Ceccato, Vania (2017), 'Women's transit safety: making connections and defining future directions in research and practice', *Crime Prevention and Community Safety*, 19:3–4 (September 2017), 276–87
62 Gardner, Cui and Coiacetto (2017)
63 http://wricitieshub.org/sites/default/files/Final_Report_24082015_0.pdf
64 Ceccato (2017)
65 https://twitter.com/awlilnatty/status/860142443550957568
66 http://www.hindustantimes.com/delhi-news/why-delhi-s-public-transport-is-still-a-war-zone-for-women/story-0bzla56HO3BIgI9LQqSSJI.html
67 Ceccato and Paz (2017)
68 http://www.slate.com/articles/double_x/doublex/2016/08/what_happens_when_sexual_assault_happens_on_a_long_haul_flight.html
69 http://www.independent.co.uk/travel/news-and-advice/woman-masturbating-passenger-cabin-crew-american-airlines-paris-a7839186.html?cmpid=facebook-post
70 Ceccato (2017)
71 http://transweb.sjsu.edu/sites/default/files/2611-women-transportation.pdf
72 Gardner, Cui and Coiacetto (2017)
73 https://matadornetwork.com/life/make-public-transportation-safer-women/
74 https://matadornetwork.com/life/make-public-transportation-safer-women/
75 http://wricitieshub.org/sites/default/files/Final_Report_24082015_0.pdf T
76 https://link.springer.com/article/10.1057/sj.2014.9; http://wricitieshub.org/sites/default/files/Final_Report_24082015_0.pdf
77 http://content.tfl.gov.uk/travel-in-london-understanding-our-diverse-communities.pdf
78 https://matadornetwork.com/life/make-public-transportation-safer-women/
79 http://news.trust.org//spotlight/most-dangerous-transport-systems-for-women/
80 http://indiatoday.intoday.in/story/delhi-gangrape-victims-friend-relives-the-horrifying-84-minutes-of-december-16-night/1/309573.html
81 https://www.nytimes.com/2017/05/05/world/asia/death-sentence-delhi-gang-rape.html
82 http://www.reuters.com/article/us-india-rape-attack-idUSBRE8BU02E20121231

83 Goodney, Suzanne, D'Silva, Lea Elsa and Asok, Abhijith (2017), 'Women's strategies addressing sexual harassment and assault on public buses: an analysis of crowdsourced data', *Crime Prevention and Community Safety*, 19: 3–4, 227–39
84 https://www.theguardian.com/global-development-professionals-network/2016/oct/13/why-arent-we-designing-cities-that-work-for-women-not-just-men
85 https://www.theguardian.com/cities/2014/dec/05/if-women-built-cities-what-would-our-urban-landscape-look-like
86 http://www.dailytitan.com/2013/11/workout-culture-subconsciously-reinforces-sexist-norms/ http://www.telegraph.co.uk/women/womens-life/11587175/Womens-fitness-What-men-really-think-about-women-in-the-gym.html
87 Irschik, Elisabeth and Kail, Eva, 'Vienna: Progress Towards a Fair Shared City', in Roberts, Marion and Sánchez de Madariaga, Inés (eds.) (2013)
88 http://www.wpsprague.com/research-1/2017/1/6/more-girls-to-parks-case-study-of-einsiedler-park-viennamilota-sidorova
89 http://civitas.eu/sites/default/files/civ_pol-an2_m_web.pdf
90 https://malmo.se/download/18.1388f79a149845ce3b9ff3/1491301765672/F%C3%B6rstudie+j%C3%A4mstalld+stadsplanering+Add+Gender+2013.pdf
91 https://malmo.se/download/18.1388f79a149845ce3b9102b/1491300931437/Presentation+20120913.pdf
92 http://webbutik.skl.se/bilder/artiklar/pdf/7164-987-4.pdf?issuusl=ignore

Chapter 3

1 https://www.theguardian.com/world/2005/oct/18/gender.uk
2 http://www.bbc.co.uk/news/magazine-34602822
3 https://eng.fjarmalaraduneyti.is/media/Gender_Equality_in_Iceland_012012.pdf
4 http://www.smh.com.au/lifestyle/health-and-wellbeing/wellbeing/what-is-life-really-like-for-women-in-iceland-the-worlds-most-womanfriendly-country-20161031-gsez8j.html
5 http://www3.weforum.org/docs/WEF_GGGR_2017.pdf
6 https://www.economist.com/blogs/graphicdetail/2016/03/daily-chart-0
7 McKinsey Global Institute (2015), *The Power of Parity: how advancing women's equality can add $12 trillion to global growth*
8 https://ourworldindata.org/women-in-the-labor-force-determinants
9 Veerle, Miranda (2011), 'Cooking, Caring and Volunteering: Unpaid Work Around the World', *OECD Social, employment and migration working papers no.116*, OECD

10 http://www.pwc.com.au/australia-in-transition/publications/understanding-the-unpaid-economy-mar17.pdf

11 Chopra, D. and Zambelli, E. (2017), 'No Time to Rest: Women's Lived Experiences of Balancing Paid Work and Unpaid Care Work', *Institute of Development Studies*

12 Veerle (2011)

13 Dinh, Huong, Strazdins, Lyndall and Welsh, Jennifer (2017), 'Hour-glass ceilings: Work-hour thresholds, gendered health inequities', *Social Science & Medicine* 176, 42–51

14 http://www.oecd.org/dev/development-gender/Unpaid_care_work.pdf

15 https://www.alzheimersresearchuk.org/wp-content/uploads/2015/03/Women-and-Dementia-A-Marginalised-Majority1.pdf

16 Ibid.

17 Ibid.

18 Ibid.

19 https://www.bls.gov/opub/ted/2015/time-spent-in-leisure-activities-in-2014-by-gender-age-and-educational-attainment.htm

20 https://www.ons.gov.uk/peoplepopulationandcommunity/wellbeing/articles/menenjoyfivehoursmoreleisuretimeperweekthanwomen/2018–01-09

21 Dinh, Strazdins and Welsh (2017)

22 http://www3.weforum.org/docs/GGGR16/WEF_Global_Gender_Gap_Report_2016.pdf

23 http://siteresources.worldbank.org/EXTSOCIALDEVELOPMENT/Resources/244362–1265299949041/6766328–1270752196897/Gender_Infrastructure2.pdf

24 L. Schiebinger and S. K. Gilmartin (2010), 'Housework is an academic issue', *Academe*, 96:39–44

25 https://www.newscientist.com/article/2085396-childcare-and-housework-are-what-give-women-more-heart-problems/

26 Kilpi, F., Konttinen, H., Silventoinen, K., Martikainen, P. (2015) 'Living arrangements as determinants of myocardial infarction incidence and survival: A prospective register study of over 300,000 Finnish men and women', *Social Science & Medicine*, 133, 93–100

27 http://www.independent.co.uk/life-style/husbands-create-extra-seven-hours-of-housework-a-week-a6885951.html

28 https://theconversation.com/census-2016-women-are-still-disadvantaged-by-the-amount-of-unpaid-housework-they-do-76008

29 https://www.inc.com/tom-popomaronis/science-says-you-shouldnt-work-more-than-this-number-of-hours-a-day.html?cid=cp01002wired

30 https://www.theguardian.com/lifeandstyle/2018/jan/15/is-28-hours-ideal-working-week-for-healthy-life

31 http://www.hse.gov.uk/statistics/causdis/stress/stress.pdf?pdf=stress

32 http://www.ilo.org/dyn/normlex/en/f?p=NORMLEXPUB:12100:0::NO::P12100_ILO_CODE:C030

33 Virtanen, M., Ferrie, J. E., Singh-Manoux, A. et al. (2011), 'Long working hours and symptoms of anxiety and depression: a 5-year follow-up of the Whitehall II study', *Psychological Medicine*, 41:12, 2485–94

34 Shields, M. (1999) 'Long working hours and health', *Health Reports*, 11:2, 33–48

35 Dinh, Strazdins and Welsh (2017)

36 Dembe, Allard E. and Yao, Xiaoxi (2016), 'Chronic Disease Risks From Exposure to Long-Hour Work Schedules Over a 32-Year Period', *MPH Journal of Occupational & Environmental Medicine*, 58:9, 861–7

37 Ibid.

38 https://www.usatoday.com/story/life/entertainthis/2017/01/08/ryan-gosling-golden-globes-acceptance-speech-eva-mendes/96330942/

39 https://www.theguardian.com/lifeandstyle/2018/mar/03/spot-working-mother-happy-busy-caretaker

40 http://www.fawcettsociety.org.uk/wp-content/uploads/2016/04/Closing-the-Pensions-Gap-Web.pdf

41 Fawcett Society (2018), *Sex Discrimination Law Review*

42 http://www.fawcettsociety.org.uk/wp-content/uploads/2016/04/Closing-the-Pensions-Gap-Web.pdf

43 https://www.closethegap.org.uk/content/gap-statistics/

44 https://www.ons.gov.uk/employmentandlabourmarket/peopleinwork/earningsandworkinghours/bulletins/annualsurveyofhoursandearnings/2017provisionaland2016revisedresults

45 https://www.statista.com/statistics/280691/median-hourly-earnings-for-part-time-employees-in-the-uk-since-2006/

46 Levanon, Asaf, England, Paula and Allison, Paul (2009) 'Occupational Feminization and Pay: Assessing Causal Dynamics Using 1950–2000 U.S. Census Data', *Social Forces*, 88:2, 865–891

47 Pan, Jessica (2015), 'Gender Segregation in Occupations: The Role of Tipping and Social Interactions', *Journal of Labor Economics*, 33:2, 365–408

48 https://www.oecd.org/dev/development-gender/Unpaid_care_work.pdf

49 Fawcett Society (2018), *Sex Discrimination Law Review*
50 Ibid.
51 http://newlaborforum.cuny.edu/2017/03/03/recognize-reduce-redistribute-unpaid-care-work-how-to-close-the-gender-gap/
52 http://progress.unwomen.org/en/2015/pdf/UNW_progressreport.pdf
53 Ibid.
54 Ibid.
55 https://www.unisa.edu.au/Global/EASS/HRI/Austen,%20Sharp%20and%20Hodgson%202015.pdf
56 http://www.fawcettsociety.org.uk/wp-content/uploads/2016/04/Closing-the-Pensions-Gap-Web.pdf
57 http://www.bbc.co.uk/news/business-39040132
58 http://www.fawcettsociety.org.uk/wp-content/uploads/2016/04/Closing-the-Pensions-Gap-Web.pdf
59 http://progress.unwomen.org/en/2015/pdf/UNW_progressreport.pdf
60 http://newlaborforum.cuny.edu/2017/03/03/recognize-reduce-redistribute-unpaid-care-work-how-to-close-the-gender-gap/
61 http://progress.unwomen.org/en/2015/pdf/UNW_progressreport.pdf
62 Kalb, Guyonne (2018), 'Paid Parental Leave and Female Labour Supply: A Review', *Economic Record*, 94:304, 80–100; Strang, Lucy and Broeks, Miriam (2016), 'Maternity leave policies: Trade-offs between labour market demands and health benefits for children', European Union; https://www.dol.gov/wb/resources/paid_parental_leave_in_the_united_states.pdf (2014)
63 Rossin-Slater, Maya, Ruhm, Christopher J. and Waldfogel, Jane (2011), 'The Effects of California's Paid Family Leave Program on Mothers' Leave-Taking and Subsequent Labor Market Outcomes', *NBER Working Paper No. 17715*; Kalb (2018)
64 Kalb (2018)
65 Strang and Broeks (2016)
66 https://www.nytimes.com/2012/08/23/technology/in-googles-inner-circle-a-falling-number-of-women.html
67 https://www.oecd.org/els/soc/PF2_1_Parental_leave_systems.pdf
68 Kalb (2018)
69 https://www.maternityaction.org.uk/2017/03/the-truth-is-that-uk-maternity-pay-is-amongst-the-lowest-in-europe/
70 https://www.oecd.org/els/soc/PF2_1_Parental_leave_systems.pdf
71 https://www.chathamhouse.org/publications/twt/brexit-isn-t-just-blokes

72 http://www.europarl.europa.eu/RegData/etudes/ATAG/2016/593543/EPRS_ATA(2016)593543_EN.pdf
73 https://politicalscrapbook.net/2017/10/mays-new-brexit-minister-wants-to-ditch-eu-laws-protecting-pregnant-women-and-vulnerable-workers/#more-67848
74 https://www.fawcettsociety.org.uk/Handlers/Download.ashx?IDMF=0de4f7f0-d1a0-4e63-94c7-5e69081caa5f
75 https://www.standard.co.uk/news/politics/councillor-dumped-from-authority-over-time-off-after-giving-birth-prematurely-10122410.html
76 https://www.weforum.org/agenda/2016/08/these-10-countries-have-the-best-parental-leave-policies-in-the-world
77 http://uk.businessinsider.com/maternity-leave-worldwide-2017–8/#us-the-family-and-medical-leave-act-provides-up-to-12-weeks-unpaid-leave-but-it-doesnt-apply-to-everyone-5
78 https://www.brookings.edu/wp-content/uploads/2017/06/es_20170606_paidfamilyleave.pdf
79 https://www.bloomberg.com/news/articles/2017–11-09/malaysia-s-giving-working-moms-a-better-maternity-deal-than-u-s
80 http://prospect.org/article/beware-paid-family-leave-fig-leaf-gop-tax-plan
81 https://www.bls.gov/ncs/ebs/benefits/2017/ebbl0061.pdf
82 https://www.independent.co.uk/news/world/americas/paid-maternity-leave-us-worst-countres-world-donald-trump-family-leave-plan-women-republican-social-a7606036.html
83 Blau, Francine D. and Kahn, Lawrence M. (2013), 'Female Labor Supply: Why is the US Falling Behind?', *The American Economic Review*, 103:3, 251–256
84 https://www.nytimes.com/2018/02/20/upshot/why-a-republican-plan-for-paid-leave-has-stirred-concern-about-social-security.html
85 http://crr.bc.edu/working-papers/how-much-does-motherhood-cost-women-in-social-security-benefits/
86 See Chapter 10
87 http://www.slate.com/blogs/xx_factor/2017/05/17/cdc_data_says_women_in_their_thirties_are_having_more_babies_than_women.html
88 https://www.theatlantic.com/sexes/archive/2013/07/for-female-scientists-theres-no-good-time-to-have-children/278165/
89 http://www.slate.com/articles/double_x/doublex/2013/06/female_academics_pay_a_heavy_baby_penalty.html
90 http://www.slate.com/articles/double_x/doublex/2013/06/female_academics_pay_a_heavy_baby_penalty.html

91 https://www.nytimes.com/2016/06/26/business/tenure-extension-policies-that-put-women-at-a-disadvantage.html

92 https://www.nytimes.com/2016/06/26/business/tenure-extension-policies-that-put-women-at-a-disadvantage.html

93 https://hardsci.wordpress.com/2016/06/28/dont-change-your-family-friendly-tenure-extension-policy-just-yet/

94 http://ec.europa.eu/eurostat/statistics-explained/images/3/39/Employment_rate_by_sex%2C_age_group_20–64%2C_1993–2016_%28%25%29.png

95 https://qz.com/266841/economic-case-for-paternity-leave/

96 https://www.oecd.org/policy-briefs/parental-leave-where-are-the-fathers.pdf

97 https://www.theguardian.com/world/2015/may/28/swedish-fathers-paid-paternity-parental-leave

98 https://www.oecd.org/policy-briefs/parental-leave-where-are-the-fathers.pdf

99 https://www.theguardian.com/money/2017/mar/22/force-men-to-take-fater-only-parental-leave-experts-urge-mps

100 https://qz.com/266841/economic-case-for-paternity-leave/

101 https://www.ft.com/content/f3154b96-e0c5-11e5-8d9b-e88a2a889797

102 https://www.nytimes.com/2014/11/09/upshot/paternity-leave-the-rewards-and-the-remaining-stigma.html

103 http://www.nytimes.com/2010/06/10/world/europe/10iht-sweden.html

104 http://fortune.com/2014/10/25/7-companies-with-the-best-perks-for-parents/

105 https://www.nytimes.com/2012/08/23/technology/in-googles-inner-circle-a-falling-number-of-women.html

106 http://www.businessinsider.com/evernote-pays-for-its-employees-to-have-their-houses-cleaned-2012–10?IR=T

107 https://www.elle.com/culture/career-politics/g28143/the-best-lactation-rooms-across-america/

108 http://time.com/money/4972232/12-companies-with-the-most-luxurious-employee-perks/

109 http://www.slate.com/blogs/xx_factor/2017/05/16/apple_s_new_headquarters_apple_park_has_no_child_care_center_despite_costing.html

110 http://www.kff.org/other/poll-finding/kaiser-family-foundationnew-york-timescbs-news-non-employed-poll/

111 https://www.flexjobs.com/blog/post/stats-about-remote-and-flexible-work-2017-predictions/

112 https://www.nbcnews.com/business/business-news/why-are-big-companies-calling-their-remote-workers-back-office-n787101
113 https://timewise.co.uk/wp-content/uploads/2017/06/Timewise-Flexible-Jobs-Index-2017.pdf
114 Goldin, Claudia (2014), 'A Grand Gender Convergence: Its Last Chapter,' *American Economic Review*, American Economic Association, 104:4, 1091–119
115 https://fivethirtyeight.com/features/why-women-are-no-longer-catching-up-to-men-on-pay/
116 European Parliament (2017), *Gender Equality and Taxation in the European Union*
117 http://www.undp.org/content/dam/undp/library/gender/Gender%20and%20Poverty%20Reduction/Taxation%20English.pdf
118 Schiebinger and Gilmartin (2010)
119 https://www.ft.com/content/60729d68-20bb-11e5-aa5a-398b2169cf79
120 http://www.economist.com/news/briefing/21599763-womens-lowly-status-japanese-workplace-has-barely-improved-decades-and-country
121 http://stats.oecd.org/index.aspx?queryid=54757
122 http://money.cnn.com/2016/10/16/news/economy/japan-companies-women-careers-nissan/index.html
123 http://www.economist.com/news/briefing/21599763-womens-lowly-status-japanese-workplace-has-barely-improved-decades-and-country
124 https://www.oecd.org/japan/japan-improving-the-labour-market-outcomes-of-women.pdf
125 https://ec.europa.eu/research/science-society/document_library/pdf_06/structural-changes-final-report_en.pdf
126 https://www.theatlantic.com/sexes/archive/2013/07/for-female-scientists-theres-no-good-time-to-have-children/278165/
127 https://work.qz.com/1156034/nobel-prize-winner-christiane-nusslein-volhard-is-helping-women-scientists-pay-to-outsource-household-chores/
128 http://genderpolicyreport.umn.edu/tax-proposals-a-missed-opportunity-for-addressing-implicit-gender-bias/; European Parliament (2017), *Gender Equality and Taxation in the European Union*
129 https://www.irs.gov/businesses/small-businesses-self-employed/deducting-business-expenses
130 http://fortune.com/2016/07/23/expense-policies-hurt-women/
131 https://www.gingerbread.org.uk/policy-campaigns/publications-index/statistics/

132 https://singlemotherguide.com/single-mother-statistics/
133 Fawcett Society (2017), *Does Local Government Work for Women?*

Chapter 4

1 Goldin, Claudia and Rouse, Cecilia (2000), 'Orchestrating Impartiality: The Impact of 'Blind' Auditions on Female Musicians', *American Economic Review*, 90:4, 715–41
2 http://www.stltoday.com/entertainment/arts-and-theatre/in-orchestras-a-sea-change-in-gender-proportions/article_25cd8c54-5ca4-529f-bb98-8c5b08c64434.html
3 https://nyphil.org/about-us/meet/musicians-of-the-orchestra
4 Kunovich, Sheri and Slomczynski, Kazimierz M. (2007), 'Systems of Distribution and a Sense of Equity: A Multilevel Analysis of Meritocratic Attitudes in Post-industrial Societies', *European Sociological Review*, 23:5, 649–63; Castilla., Emilio J. and Benard, Stephen (2010), 'The Paradox of Meritocracy in Organizations', *Administrative Science Quarterly*, 55:4, 543–676
5 Reynolds, Jeremy and Xian, He (2014), 'Perceptions of meritocracy in the land of opportunity', *Research in Social Stratification and Mobility*, 36, 121–37
6 Castilla and Benard (2010)
7 http://fortune.com/2014/08/26/performance-review-gender-bias/
8 Castilla and Benard (2010)
9 http://stateofstartups.firstround.com/2016/#highlights-diversity-prediction
10 Uhlmann, Eric Luis and Cohen, Geoffrey L. (2007), '"I think it, therefore it's true": Effects of self-perceived objectivity on hiring discrimination', *Organizational Behavior and Human Decision Processes*, 104:2, 207–23; Castilla and Benard (2010)
11 https://www.newyorker.com/magazine/2017/11/20/the-tech-industrys-gender-discrimination-problem
12 https://www.newyorker.com/magazine/2017/11/20/the-tech-industrys-gender-discrimination-problem
13 https://hbr.org/2014/10/hacking-techs-diversity-problem
14 https://www.theatlantic.com/magazine/archive/2017/04/why-is-silicon-valley-so-awful-to-women/517788/
15 http://www.latimes.com/business/la-fi-women-tech-20150222-story.html#page=1
16 Reynolds and Xian (2014)

17 Handley, Ian M., Brown, Elizabeth R., Moss-Racusin, Corinne A. and Smith, Jessi L. (2015), 'Quality of evidence revealing subtle gender biases in science is in the eye of the beholder', *Proceedings of the National Academy of Sciences of the United States of America*, 112:43, 13201–13206

18 https://erc.europa.eu/sites/default/files/document/file/Gender_statistics_April_2014.pdf; Wenneras, C. and Wold, A. (1997), 'Nepotism and sexism in peer-review', *Nature*, 387:341; Milkman, Katherine L., Akinola, Modupe and Chugh, Dolly (2015), 'What Happens Before? A Field Experiment Exploring How Pay and Representation Differentially Shape Bias on the Pathway Into Organizations', *Journal of Applied Psychology*, 100:6, 1678–712; Knobloch-Westerwick, Silvia, Glynn, Carroll J. and Huge, Michael (2013), 'The Matilda Effect in Science Communication', *Science Communication*, 35:5, 603–25; Kaatz, Anna, Gutierrez, Belinda and Carnes, Molly (2014), 'Threats to objectivity in peer review: the case of gender', *Trends Pharmacol Sci.*, 35:8, 371–3; Women and Science Unit (2011), *White Paper on the Position of Women in Science in Spain*, UMYC

19 Women and Science Unit (2011); https://foreignpolicy.com/2016/04/19/how-to-get-tenure-if-youre-a-woman-academia-stephen-walt/

20 Roberts, Sean G. and Verhoef, Tessa (2016), 'Double-blind reviewing at EvoLang 11 reveals gender bias', *Journal of Language Evolution*, 1:2, 163–67

21 Budden, Amber E., Tregenza, Tom, Aarssen, Lonnie W., Koricheva, Julia, Leimu, Roosa and Lortie, Christopher J. (2008), 'Double-blind review favours increased representation of female authors', *Trends in Ecology & Evolution*, 23:1, 4–6

22 Knobloch-Westerwick, Glynn and Huge (2013); Maliniak, Daniel, Powers, Ryan and Walter, Barbara F. (2013), 'The Gender Citation Gap in International Relations', International Organization; Mitchell, Sara McLaughlin, Lange, Samantha and Brus, Holly (2013), 'Gendered Citation Patterns in International Relations', *Journal of International Studies Perspectives*, 14:4, 485–92

23 King, Molly M., Bergstrom, Carl T., Correll, Shelley J., Jacquet, Jennifer, and West, Jevin D. (2017), 'Men Set Their Own Cites High: Gender and Self-citation across Fields and over Time' *Socius: Sociological Research for a Dynamic World*, 3: 1–22

24 Bagilhole, Barbara and Goode, Jackie (2001), 'The Contradiction of the Myth of Individual Merit, and the Reality of a Patriarchal Support System in Academic Careers: A Feminist Investigation', *European Journal of Women's Studies*, 8:2, 161–80

25 Krawczyk, Michał (2017), Are all researchers male? Gender misattributions in citations, *Scientometrics*, 110:3, 1397–1402

26 https://www.nytimes.com/2015/11/12/upshot/even-famous-female-economists-get-no-respect.html

27 https://www.nytimes.com/2016/01/10/upshot/when-teamwork-doesnt-work-for-women.html?mcubz=1

28 Knobloch-Westerwick, Glynn and Huge (2013)

29 https://foreignpolicy.com/2016/04/19/how-to-get-tenure-if-youre-a-woman-academia-stephen-walt/

30 https://www.chronicle.com/article/Thanks-for-Listening/233825

31 http://www.cbc.ca/news/canada/british-columbia/female-profs-more-work-1.4473910

32 Mitchell, Sara McLaughlin and Hesli, Vicki L., 'Women Don't Ask? Women Don't Say No? Bargaining and Service in the Political Science Profession, *PS: Political Science & Politics*, 46:2, 355–369; Guarino, Cassandra M. and Borden, Victor M. H. (2017), 'Faculty Service Loads and Gender: Are Women Taking Care of the Academic Family?', *Research in Higher Education*, 58:6 672–694

33 https://hbr.org/2018/03/for-women-and-minorities-to-get-ahead-managers-must-assign-work-fairly; Laperrière, Ève, Messing, Karen and Bourbonnais, Renée (2017), 'Work activity in food service: The significance of customer relations, tipping practices and gender for preventing musculoskeletal disorders', *Applied Ergonomics*, 58, 89–101

34 Guarino and Borden (2017); Baker, Maureen (2012), *Academic Careers and the Gender Gap*, Canada; Gibney, Elizabeth (2017), 'Teaching load could put female scientists at career disadvantage', *Nature*, https://www.nature.com/news/teaching-load-could-put-female-scientists-at-career-disadvantage-1.21839; Women and Science Unit (2011), *White Paper on the Position of Women in Science in Spain*, UMYC

35 Amy Bug (2010), 'Swimming against the unseen tide', *Phys. World*, 23:08; Boring, Anne, Ottoboni, Kellie and Stark, Philip B. (2016), 'Student evaluations of teaching (mostly) do not measure teaching effectiveness' *ScienceOpen Research*

36 Boring, Anne, Ottoboni, Kellie and Stark, Philip B. (2016)

37 http://activehistory.ca/2017/03/shes-hot-female-sessional-instructors-gender-bias-andf-student-evaluations/

38 MacNell, Lillian, Driscoll, Adam and Hunt, Andrea N. (2015), 'What's in a Name: Exposing Gender Bias in Student Ratings of Teaching', *Innovative Higher Education*, 40:4, 291–303

39 https://www.theguardian.com/lifeandstyle/womens-blog/2015/feb/13/female-academics-huge-sexist-bias-students

40 http://activehistory.ca/2017/03/shes-hot-female-sessional-instructors-gender-bias-and-student-evaluations/

41 Storage, Daniel, Home, Zachary, Cimpian, Andrei and Leslie, Sarah-Jane (2016), 'The Frequency of "Brilliant" and "Genius" in Teaching Evaluations Predicts the Representation of Women and African Americans across Fields', *PLoS ONE* 11:3; Leslie, Sarah-Jane, Cimpian, Andrei, Meyer, Meredith and Freeland, Edward (2015), 'Expectations of brilliance underlie gender distributions across academic disciplines', *Science*, 347:6219, 262–5; Meyer, Meredith, Cimpian, Andrei and Leslie, Sarah-Jane (2015), 'Women are underrepresented in fields where success is believed to require brilliance', *Frontiers in Psychology*, 6:235

42 Banchefsky, Sarah, Westfall, Jacob, Park, Bernadette and Judd, Charles M. (2016), 'But You Don't Look Like A Scientist!: Women Scientists with Feminine Appearance are Deemed Less Likely to be Scientists', *Sex Roles*, 75:3–4, 95–109

43 Bian, Lin, Leslie, Sarah-Jane and Cimpian, Andrei (2017), 'Gender stereotypes about intellectual ability emerge early and influence children's interests', *Science*, 355: 6323, 389–391

44 https://genderedinnovations.stanford.edu/institutions/bias.html

45 https://www.theguardian.com/commentisfree/2016/may/31/women-science-industry-structure-sexist-courses-careers

46 Grunspan, Daniel Z., Eddy, Sarah L., Brownell, Sara E., Wiggins, Benjamin L., Crowe, Alison J., Goodreau, Steven M. (2016), 'Males Under-Estimate Academic Performance of Their Female Peers in Undergraduate Biology Classrooms', *PLoS ONE*, 11:2

47 Schmader, Toni, Whitehead, Jessica and Wysocki, Vicki H. (2007), 'A Linguistic Comparison of Letters of Recommendation for Male and Female Chemistry and Biochemistry Job Applicants', *Sex Roles*, 57:7–8, 509–14; Madera, Juan M., Hebl, Michelle R. and Martin, Randi C. (2009), 'Gender and letters of recommendation for academia: Agentic and communal differences', *Journal of Applied Psychology*, 94:6, 1591–9; Dutt, Kuheli, Pfaff, Danielle L., Bernstein, Ariel F., Dillard, Joseph S . and Block, Caryn J. (2016), 'Gender differences in recommendation letters for postdoctoral fellowships in geoscience', *Nature Geoscience*, 9, 805–8

48 Madera et al. (2009)

49 https://www.nature.com/news/women-postdocs-less-likely-than-men-to-get-a-glowing-reference-1.20715

50 Trix, Frances and Psenka, Carolyn (2003), 'Exploring the Color of Glass: Letters of Recommendation for Female and Male Medical Faculty', *Discourse & Society*, 14:2, 191–220

51 Ibid.

52 Madera at al. (2009)

53 Nielsen, Mathias Wullum, Andersen, Jens Peter, Schiebinger, Londa and Schneider, Jesper W. (2017), 'One and a half million medical papers reveal a link between author gender and attention to gender and sex analysis', *Nature Human Behaviour*, 1, 791–6

54 http://gap.hks.harvard.edu/effects-gender-stereotypic-and-counter-stereotypic-textbook-images-science-performance

55 https://www.cs.cmu.edu/afs/cs/project/gendergap/www/papers/anatomy-WSQ99.html

56 Light, Jennifer S. (1999), 'When Computers Were Women', *Technology and Culture*, 40:3, 455–483

57 Ensmenger, Nathan L. (2010), *The Computer Boys Take Over: Computers, Programmers, and the Politics of Technical Expertise*, Cambridge MA

58 https://www.theatlantic.com/business/archive/2016/09/what-programmings-past-reveals-about-todays-gender-pay-gap/498797/

59 http://thecomputerboys.com/wp-content/uploads/2011/06/cosmopolitan-april-1967–1-large.jpg

60 https://www.theatlantic.com/business/archive/2016/09/what-programmings-past-reveals-about-todays-gender-pay-gap/498797/

61 Ensmenger, Nathan L. (2010)

62 Ibid.

63 https://www.hfobserver.com/exclusive-content/q4-top-recruiting-department-hires-and-an-acquisition/

64 https://www.theguardian.com/science/2016/sep/01/how-algorithms-rule-our-working-lives

65 https://www.theatlantic.com/technology/archive/2013/11/your-job-their-data-the-most-important-untold-story-about-the-future/281733/

66 https://onlinelibrary.wiley.com/doi/abs/10.1111/j.1471–6402.2008.00454.x; Hannah Riley Bowles, Linda Babcock and Lei Lai (2007), 'Social incentives for gender differences in the propensity to initiate negotiations: Sometimes it does hurt to ask', *Organizational Behavior and Human Decision Processes*, 103, 84–103.

67 https://www.nytimes.com/2012/08/23/technology/in-googles-inner-circle-a-falling-number-of-women.html

68 https://www.physiology.org/doi/10.1152/advan.00085.2017

69 https://www.newyorker.com/magazine/2017/11/20/the-tech-industrys-gender-discrimination-problem

70 https://medium.com/@triketora/where-are-the-numbers-cb997a57252

71 http://www.independent.co.uk/news/business/news/workplace-gender-quotas-incompetence-efficiency-business-organisations-london-school-economics-lse-a7797061.html

72 http://web.mit.edu/fnl/volume/184/hopkins.html

73 http://www.cwf.ch/uploads/press/ABusinessCaseForWomen.pdf

74 https://madebymany.com/stories/can-a-few-well-chosen-words-improve-inclusivity

75 Gaucher, D., Friesen, J. and Kay, A. C. (2011), 'Evidence that gendered wording in job advertisements exists and sustains gender inequality', *Journal of Personality and Social Psychology*, 101:1, 109–28

76 https://www.theatlantic.com/business/archive/2015/12/meritocracy/418074/

77 Castilla, Emilio J. (2015), 'Accounting for the Gap: A Firm Study Manipulating Organizational Accountability and Transparency in Pay Decisions', *Organization Science*, 26:2, 311–33

Chapter 5

1 Kingma, Boris and Marken Lichtenbelt, Wouter van (2015), 'Energy consumption in buildings and female thermal demand,' *Nature Climate Change*, 5, 1054–6

2 https://www.nytimes.com/2015/08/04/science/chilly-at-work-a-decades-old-formula-may-be-to-blame.html?_r=0

3 http://www.hse.gov.uk/statistics/history/historical-picture.pdf

4 http://www.hse.gov.uk/statistics/pdf/fatalinjuries.pdf

5 https://www.cdc.gov/mmwr/preview/mmwrhtml/mm4822a1.htm

6 https://www.bls.gov/news.release/cfoi.nr0.htm

7 https://www.equaltimes.org/the-invisible-risks-facing-working?lang=en#.W0oUw9gzrOT

8 Ibid

9 http://www.hazards.org/vulnerableworkers/ituc28april.htm

10 https://www.equaltimes.org/the-invisible-risks-facing-working?lang=en#.WsyCV9MbPOS

11 Messing, K. (in press), 'Fighting invisibility in the workplace: the struggle to protect health and support equality in the workplace' In Greaves, Lorraine (ed.) *A History of Women's Health in Canada*, Second Story Press.

12 Côté, Julie (2012), 'A critical review on physical factors and functional characteristics that may explain a sex/gender difference in work-related neck/shoulder disorders', *Ergonomics*, 55:2, 173–82
13 http://www.hse.gov.uk/statistics/causdis/cancer/cancer.pdf?pdf=cancer
14 Rochon Ford, Anne (2014), "Overexposed, Underinformed": Nail Salon Workers and Hazards to Their Health / A Review of the Literature National Network on Environments and Women's Health', RPSFM (Réseau pancanadien sur la santé des femmes et le milieu)
15 http://www.hazards.org/vulnerableworkers/ituc28april.htm
16 'Breast Cancer and Occupation: The Need for Action: APHA Policy Statement Number 20146, Issued November 18, 2014', *NEW SOLUTIONS: A Journal of Environmental and Occupational Health Policy*; Rochon Ford (2014)
17 'Breast Cancer and Occupation: The Need for Action: APHA Policy Statement Number 20146, Issued November 18, 2014'; Brophy, James T., Keith, Margaret M. et al. (2012), 'Breast cancer risk in relation to occupations with exposure to carcinogens and endocrine disruptors: a Canadian case-control study', *Environmental Health*, 11:87
18 Rochon Ford (2014)
19 http://www.passblue.com/2017/07/05/females-exposed-to-nuclear-radiation-are-far-likelier-than-males-to-suffer-harm/
20 Phillips, Ann M. (2014), 'Wonderings on Pollution and Women's Health', in Scott, Dayna Nadine (ed.), *Our Chemical Selves: Gender, Toxics, and Environmental Health*, Vancouver
21 Scott, Dayna Nadine and Lewis, Sarah (2014), 'Sex and Gender in Canada's Chemicals Management Plan', in Scott, Dayna Nadine (ed.), *Our Chemical Selves: Gender, Toxics, and Environmental Health*, Vancouver
22 Rochon Ford (2014)
23 Scott and Lewis (2014)
24 Rochon Ford (2014)
25 Scott and Lewis (2014)
26 Ibid.
27 Rochon Ford (2014)
28 Scott and Lewis (2014)
29 'Breast Cancer and Occupation: The Need for Action: APHA Policy Statement Number 20146, Issued November 18, 2014', *NEW SOLUTIONS: A Journal of Environmental and Occupational Health Policy*
30 Rochon Ford (2014)
31 Brophy et al. (2012)

32 'Breast Cancer and Occupation: The Need for Action: APHA Policy Statement Number 20146, Issued November 18, 2014', *NEW SOLUTIONS: A Journal of Environmental and Occupational Health Policy*
33 https://www.theguardian.com/lifeandstyle/2015/may/05/osha-health-women-breast-cancer-chemicals-work-safety
34 https://www.theguardian.com/lifeandstyle/2015/apr/30/fda-cosmetics-health-nih-epa-environmental-working-group
35 Rochon Ford (2014); Brophy et al. (2012); Scott and Lewis (2014)
36 Scott and Lewis (2014)
37 Brophy et al. (2012)
38 Scott and Lewis (2014)
39 http://www.hazards.org/compensation/meantest.htm
40 'Designing Tools and Agricultural Equipment for Women', poster produced by Aaron M. Yoder, Ann M. Adams and Elizabeth A. Brensinger, for 2014 Women in Agriculture Educators National Conference
41 http://nycosh.org/wp-content/uploads/2014/09/Women-in-Construction-final-11–8-13–2.pdf
42 Myles, Kimberly and Binseel, Mary S. (2007), 'The Tactile Modality: A Review of Tactile Sensitivity and Human Tactile Interfaces', Army Research Laboratory
43 http://www.afpc.af.mil/About/Air-Force-Demographics/
44 https://www.gov.uk/government/uploads/system/uploads/attachment_data/file/389575/20141218_WGCC_Findings_Paper_Final.pdf
45 https://www.theguardian.com/uk-news/2013/nov/24/female-raf-recruits-compensation-marching-injuries
46 Laperrière, Ève, Messing, Karen and Bourbonnais, Renée (2017), 'Work activity in food service: The significance of customer relations, tipping practices and gender for preventing musculoskeletal disorders', *Applied Ergonomics*, 58, 89–101
47 Friedl, Karl E. (2012), 'Military Quantitative Physiology: Problems and Concepts in Military Operational Medicine', Office of the Surgeon General, Department of the Army, United States of America; Knapik, Joseph and Reynolds, Katy (2012), 'Load Carriage in Military Operations A Review of Historical, Physiological, Biomechanical, and Medical Aspects', Walter Reed Army Medical Center, US Army Medical Department Center & School
48 https://assets.publishing.service.gov.uk/government/uploads/system/uploads/attachment_data/file/389575/20141218_WGCC_Findings_Paper_Final.pdf
49 Ibid.

50 ibid.

51 http://www.independent.co.uk/news/world/americas/dressed-to-kill-us-army-finally-designs-a-female-uniform-that-fits-2274446.html

52 https://www.washingtontimes.com/news/2015/may/14/military-pressed-to-design-line-of-women-friendly-/

53 https://www.tuc.org.uk/sites/default/files/PPEandwomenguidance.pdf

54 https://blogs.scientificamerican.com/voices/one-more-barrier-faced-by-women-in-science/

55 https://www.tuc.org.uk/sites/default/files/PPEandwomenguidance.pdf

56 https://www.wes.org.uk/sites/default/files/WES%20safety%20survey%20results%20March%202010.pdf

57 www.prospect.org.uk/news/id/2016/June/21/Women-workers-highlight-problems-with-ill-fitting-protective-equipment

58 https://www.tuc.org.uk/sites/default/files/2016–01299-Leaflet-booklet-Women%27s-PPE–One-Size-Does-Not-Fit-All-Version-26–09-2016%20%282%29.pdf

59 http://nycosh.org/wp-content/uploads/2014/09/Women-in-Construction-final-11–8-13–2.pdf

60 http:www.prospect.org.uk/news/id/2016/June/21/Women-workers-highlight-problems-with-ill-fitting-protective-equipment

61 https://www.tuc.org.uk/sites/default/files/PPEandwomenguidance.pdf

62 theguardian.com/world/2016/sep/25/spain-guardia-civil-sexism-women-bulletproof-jackets?client=safari

63 theguardian.com/world/2016/sep/25/spain-guardia-civil-sexism-women-bulletproof-jackets?client=safari

64 https://www.tuc.org.uk/sites/default/files/PPEandwomenguidance.pdf

Chapter 6

1 Vogel, Sarah A. (2009), The Politics of Plastics: The Making and Unmaking of Bisphenol A 'Safety', *American Journal of Public Health.* 99:3, 559–566

2 http://www.washingtonpost.com/wp-dyn/content/article/2008/04/15/AR2008041501753.html

3 Vogel, Sarah A. (2009)

4 Ibid.

5 Ibid.

6 https://www.nytimes.com/2015/05/11/nyregion/nail-salon-workers-in-nyc-face-hazardous-chemicals.html; 'Breast Cancer and Occupation: The Need for Action: APHA Policy Statement Number 20146, Issued November

18, 2014', *NEW SOLUTIONS: A Journal of Environmental and Occupational Health Policy*

7 Vogel, Sarah A. (2009)

8 https://www.nytimes.com/2015/05/10/nyregion/at-nail-salons-in-nyc-manicurists-are-underpaid-and-unprotected.html

9 https://www.nytimes.com/2015/05/10/nyregion/at-nail-salons-in-nyc-manicurists-are-underpaid-and-unprotected.html

10 https://www.theguardian.com/world/2017/sep/11/slavery-report-sounds-alarm-over-vietnamese-nail-bar-workers

11 https://www.theguardian.com/commentisfree/2018/jan/05/nail-bars-modern-slavery-discount-salons-booming-exploitation

12 https://www.theguardian.com/world/2017/sep/11/slavery-report-sounds-alarm-over-vietnamese-nail-bar-workers

13 https://www.tuc.org.uk/sites/default/files/the-gig-is-up.pdf

14 https://www.tuc.org.uk/sites/default/files/Women_and_casualisation_0.pdf

15 https://www.ituc-csi.org/IMG/pdf/Women_8_march_EN.pdf

16 https://www.unison.org.uk/content/uploads/2014/06/On-line-Catalogue224222.pdf

17 https://www.unison.org.uk/content/uploads/2014/06/On-line-Catalogue224222.pdf

18 http://survation.com/women-on-low-paid-zero-hours-contracts-survation-for-fawcett-society/

19 https://www.tuc.org.uk/sites/default/files/the-gig-is-up.pdf

20 http://www.ucu.org.uk/media/6882/Zero-hours-contracts-a-UCU-briefing-Mar-14/pdf/ucu_zerohoursbriefing_mar14.pdf

21 https://www.hesa.ac.uk/files/pre-release/staff_1516_table_B.xlsx

22 Best, Kathinka, Sinell, Anna, Heidingsfelder, Marie Lena and Schraudner, Martina (2016), 'The gender dimension in knowledge and technology transfer – the German case', *European Journal of Innovation Management*, 19:1, 2–25

23 A. Hellum and H. Aasen (eds.) (2013), *Women's Human Rights: CEDAW in International, Regional and National Law (Studies on Human Rights Conventions)* (Cambridge University Press, Cambridge)

24 https://www.oecd.org/japan/japan-improving-the-labour-market-outcomes-of-women.pdf

25 https://krueger.princeton.edu/sites/default/files/akrueger/files/katz_krueger_cws_-_march_29_20165.pdf

26 TUC (2017), 'The gig is up', https://www.tuc.org.uk/sites/default/files/the-gig-is-up.pdf

27 Rubery, Jill, Grimshaw, Damian and Figueiredo, Hugo (2005), 'How to close the gender pay gap in Europe: towards the gender mainstreaming of pay policy', *Industrial Relations Journal*, 36:3, 184–213

28 https://www.tuc.org.uk/sites/default/files/Women_and_casualisation.pdf

29 Ibid.

30 Ibid.

31 Ibid.

32 https://www.nytimes.com/interactive/2014/08/13/us/starbucks-workers-scheduling-hours.html

33 Ibid.

34 https://www.brookings.edu/wp-content/uploads/2017/10/es_121917_the51percent_ebook.pdf

35 https://www.ituc-csi.org/IMG/pdf/women.pdf; https://publications.parliament.uk/pa/cm201719/cmselect/cmwomeq/725/72504.htm

36 https://www.tuc.org.uk/sites/default/files/Women_and_casualisation.pdf

37 http://endviolence.un.org/pdf/pressmaterials/unite_the_situation_en.pdf

38 http://www.scmp.com/news/china/society/article/2054525/young-chinese-women-dare-say-no-workplace-sexual-harassment-says

39 https://www.ncbi.nlm.nih.gov/pubmed/19862867

40 https://www.tuc.org.uk/sites/default/files/SexualHarassmentreport2016.pdf

41 https://www.elephantinthevalley.com/

42 Brophy, James T., Keith, Margaret M. and Hurley, Michael (2018), 'Assaulted and Unheard: Violence Against Healthcare Staff', *NEW SOLUTIONS: A Journal of Environmental and Occupational Health Policy*, 27:4, 581–606

43 Ibid.

44 https://www.tuc.org.uk/sites/default/files/SexualHarassmentreport2016.pdf; https://qz.com/931653/indias-long-history-with-sexual-harassment-at-workplaces/; http://economictimes.indiatimes.com/magazines/panache/predators-at-the-workplace-india-inc-yet-to-commit-to-law-against-sexual-harassment/articleshow/57830600.cms; http://indianexpress.com/article/india/38-per-cent-women-say-they-faced-sexual-harassment-at-workplace-survey-4459402/; https://today.yougov.com/news/2017/04/25/nearly-third-women-have-been-sexually-harassed-work/; https://www.theguardian.com/money/2016/jul/22/sexual-harassment-at-work-roger-ailes-fox-news; https://www.elephantinthevalley.com/; https://interagencystandingcommittee.org/system/files/hwn_full_survey_results_may_2016.pdf

45 https://www.theguardian.com/money/2016/jul/22/sexual-harassment-at-work-roger-ailes-fox-news; https://www.elephantinthevalley.com/; https://interagencystandingcommittee.org/system/files/hwn_full_survey_results_may_2016.pdf;

46 https://www.tuc.org.uk/sites/default/files/SexualHarassmentreport2016.pdf

47 https://hbr.org/2014/10/hacking-techs-diversity-problem; https://hbr.org/2008/06/stopping-the-exodus-of-women-in-science

Chapter 7

1 https://blog.oup.com/2013/06/agriculture-gender-roles-norms-society/

2 Ibid.

3 Sağiroğlu, İsa, Kurt, Cem, Ömürlü, İmran Kurt and Çatikkaş, Fatih (2017), 'Does Hand Grip Strength Change With Gender? The Traditional Method vs. the Allometric Normalisation Method', *European Journal of Physical Education and Sports Science*, 2:6, 84–93

4 Leyk, D., Gorges, W., Ridder, D., Wunderlich, M., Ruther, T., Sievert, A. and Essfeld, D. (2007), 'Hand-grip strength of young men, women and highly trained female athletes', *European Journal of Applied Physiology*, 99, 415–21

5 Lewis, D. A., Kamon, E. and Hodgson, J. L. (1986), 'Physiological Differences Between Genders Implications for Sports Conditioning', *Sports Medicine*, 3, 357–69; Rice, Valerie J. B., Sharp, Marilyn A., Tharion, William J. and Williamson, Tania L. (1996), 'The effects of gender, team size, and a shoulder harness on a stretcher-carry task and post-carry performance. Part II. A mass-casualty simulation', *International Journal of Industrial Ergonomics*, 18, 41–9; Miller, A. E., MacDougall, J. D., Tarnopolsky, M. A. and Sale, D. G. (1993), 'Gender differences in strength and muscle fiber characteristic', *European Journal of Applied Physiology*, 66:3, 254–62

6 Lewis et al. (1986)

7 Zellers, Kerith K. and Hallbeck, M. Susan (1995), 'The Effects of Gender, Wrist and Forearm Position on Maximum Isometric Power Grasp Force, Wrist Force, and their Interactions', *Proceedings of the Human Factors and Ergonomics Society Annual Meeting*, 39:10, 543–7; Bishu, Ram R., Bronkema, Lisa A, Garcia, Dishayne, Klute, Glenn and Rajulu, Sudhakar (1994), 'Tactility as a function of Grasp force: effects of glove, orientation, pressure, load and handle', *NASA technical paper 3474* May 1994; Puh, Urška (2010), 'Age-related and sex-related differences in hand and pinch grip strength in adults', *International Journal of Rehabilitation Research*, 33:1

8 https://www.ft.com/content/1d73695a-266b-11e6-8b18-91555f2f4fde

9 Leyk et al. (2007)

10 Alesina, Alberto F., Giuliano, Paola and Nunn, Nathan (2011), 'On the Origins of Gender Roles: Women and the Plough', *Working Paper 17098*, National Bureau of Economic Research (May 2011)

11 Ibid.

12 Gella, A. A., Tadele, Getnet (2014), 'Gender and farming in Ethiopia: an exploration of discourses and implications for policy and research', FAC Working Paper, *Future Agricultures*, 84:15

13 http://www.fao.org/3/a-am309e.pdf; http://www.greenpeace.org/international/en/news/Blogs/makingwaves/international-womens-day-2017-change/blog/58902/; https://www.theguardian.com/global-development/2014/oct/16/world-food-day-10-myths-hunger; http://cmsdata.iucn.org/downloads/climate_change_gender.pdf

14 Doss, Cheryl (2011), 'If women hold up half the sky, how much of the world's food do they produce?', *ESA Working Paper No. 11*, Agricultural Development Economics Division, FAO

15 World Bank (2014), 'Levelling the Field: Improving Opportunities for Women Farmers in Africa'

16 https://openknowledge.worldbank.org/bitstream/handle/10986/15577/wps6436.pdf?sequence=1&isAllowed=y

17 Ibid.

18 https://www.theguardian.com/global-development/2016/apr/07/leaving-women-girls-out-of-development-statistics-doesnt-add-up

19 Doss (2011)

20 Petrics, H. et al. (2015), 'Enhancing the potential of family farming for poverty reduction and food security through gender-sensitive rural advisory services', UN/FAO

21 Ibid.

22 Ibid.

23 Ibid.

24 https://www.gatesfoundation.org/What-We-Do/Global-Development/Agricultural-Development/Creating-Gender-Responsive-Agricultural-Development-Programs We receive some grant proposals that do not account for gender differences and do not consider how agricultural initiatives may benefit or hinder women or men.

25 http://data2x.org/wp-content/uploads/2014/08/What-Is-Wrong-with-Data-on-Women-and-Girls_November-2015_WEB_1.pdf

26 Petrics et al. (2015)

27 http://people.brandeis.edu/~nmenon/Draft04_Womens_Empowerment_and_Economic_Development.pdf
28 https://docs.gatesfoundation.org/documents/gender-responsive-orientation-document.pdf
29 Doss (2011)
30 http://www.poverty-action.org/study/demand-nontraditional-cook stoves-bangladesh
31 http://greenwatchbd.com/70000-improved-stoves-distributed-to-combat-indoor-pollution/
32 http://www.sciencedirect.com/science/article/pii/S0160412016307358
33 Crewe, Emma et al. (2015), 'Building a Better Stove: The Sri Lanka Experience', Practical Action, Sri Lanka
34 http://www.sciencedirect.com/science/article/pii/S0160412016307358
35 Ibid.
36 Ibid.; https://www.unicef.org/health/files/health_africamalaria.pdf
37 http://greenwatchbd.com/70000-improved-stoves-distributed-to-combat-indoor-pollution/
38 https://www.unicef.org/environment/files/Bangladesh_Case_Study_2014.pdf
39 http://www.unwomen.org/en/news/stories/2012/4/green-cook-stoves-improving-women-s-lives-in-ghana#sthash.IZM4RsCG.dpuf
40 http://www.sciencedirect.com/science/article/pii/S0160412016307358
41 Crewe, Emma (1997), 'The Silent Traditions of Developing Cooks', in R. D. Grillo and R. L. Stirrat (eds.), *Discourses of Development*, Oxford
42 Ibid.
43 http://www.gender-summit.eu/images/Reports/Gender_and_inclusive_innovation_Gender_Summit_report.pdf
44 http://www.unwomen.org/-/media/headquarters/attachments/sections/library/publications/2014/unwomen_surveyreport_advance_16oct.pdf?vs=2710
45 Fatema, Naureen (2005), 'The Impact of Structural Gender Differences and its Consequences on Access to Energy in Rural Bangladesh', Asia Sustainable and Alternative Energy Program (ASTAE), Energy Wing of the World Bank Group
46 Crewe (1997), in Grillo and Stirrat (eds.)
47 http://www.washplus.org/sites/default/files/bangladesh-consumer_preference2013.pdf
48 http://answers.practicalaction.org/our-resources/item/building-a-better-stove-the-sri-lanka-experience#

49 Crewe (1997), in Grillo and Stirrat (eds.)
50 http://www.ideasrilanka.org/PDFDownloads/Cook%20Stoves%20in%20Sri%20LAnka.pdf
51 http://www.gender-summit.eu/images/Reports/Gender_and_inclusive_innovation_Gender_Summit_report.pdf
52 http://www.gender-summit.eu/images/Reports/Gender_and_inclusive_innovation_Gender_Summit_report.pdf
53 http://www.poverty-action.org/study/demand-nontraditional-cookstoves-bangladesh
54 https://www.se4all-africa.org/fileadmin/uploads/se4all/Documents/Abidjan_workshop_2016/SE_gender_GACC.PDF
55 https://news.yale.edu/2012/06/29/despite-efforts-change-bangladeshi-women-prefer-use-pollution-causing-cookstoves
56 Petrics et al. (2015)
57 http://answers.practicalaction.org/our-resources/item/building-a-better-stove-the-sri-lanka-experience#
58 https://www.se4all-africa.org/fileadmin/uploads/se4all/Documents/Abidjan_workshop_2016/SE_gender_GACC.PDF
59 https://www.thesolutionsjournal.com/article/how-a-simple-inexpensive-device-makes-a-three-stone-hearth-as-efficient-as-an-improved-cookstove/
60 Parigi, Fabio, Viscio, Michele Del, Amicabile, Simone, Testi, Matteo, Rao, Sailesh, Udaykumar, H. S. (2016), 'High efficient Mewar Angithi stove testing in rural Kenya', 7th International Renewable Energy Congress (IREC)
61 http://www.green.it/mewar-angithi/

Chapter 8

1 https://www.ncbi.nlm.nih.gov/pubmed/5550584; http://www.who.int/gender/documents/Genderworkhealth.pdf; Boyle, Rhonda and Boyle, Robin (2009), 'Hand Size and the Piano Keyboard. Literature Review and aSurvey of the Technical and Musical Benefits for Pianists using Reduced-Size Keyboards in North America', 9th Australasian Piano Pedagogy Conference, Sydney; Boyle, Rhonda, Boyle, Robin and Booker, Erica (2015), 'Pianist Hand Spans: Gender and Ethnic Differences and Implications For Piano Playing, 15th Australasian Piano Pedagogy Conference, Melbourne 2015
2 Boyle, Boyle and Booker (2015); Boyle and Boyle (2009)

3 Boyle, Boyle and Booker (2015)
4 Ibid.
5 Ibid.
6 http://www.smallpianokeyboards.org/hand-span-data.html
7 'Small hands? Try this Keyboard, You'll Like It', *Piano & Keyboard Magazine* (July/August 1998)
8 Boyle, Boyle and Booker (2015)
9 https://deviceatlas.com/blog/most-popular-smartphone-screen-sizes-2017
10 http://www.telegraph.co.uk/technology/apple/iphone/11335574/Women-more-likely-to-own-an-iPhone-than-men.html
11 https://medium.com/technology-and-society/its-a-mans-phone-a26c6-bee1b69#.mk7sjtewi
12 http://www.sciencedirect.com/science/article/pii/S1013702515300270
13 http://www.sciencedirect.com/science/article/pii/S1050641108001909
14 http://www.sciencedirect.com/science/article/pii/S1013702515300270
15 http://ac.els-cdn.com/S0169814115300512/1-s2.0-S0169814115300512-main.pdf?_tid=4235fa34-f81e-11e6-a430-00000aab0f26&acdnat=1487672132_c2148a0040def1129abc7acffe03e57d
16 Ibid.
17 http://www.sciencedirect.com/science/article/pii/S0169814116300646; http://ac.els-cdn.com/S0003687011000962/1-s2.0-S0003687011000962-main.pdf?_tid=f0a12b58-f81d-11e6-af6b-00000aab0f26&acdnat=1487671995_41cfe19ea98e87fb7e3e693bdddaba6e; http://www.sciencedirect.com/science/article/pii/S1050641108001909
18 https://www.theverge.com/circuitbreaker/2016/7/14/12187580/keeco-k1-hexagon-phone-for-women
19 https://www.theguardian.com/technology/askjack/2016/apr/21/can-speech-recognition-software-help-prevent-rsi
20 https://makingnoiseandhearingthings.com/2016/07/12/googles-speech-recognition-has-a-gender-bias/
21 http://blog-archive.griddynamics.com/2016/01/automatic-speech-recognition-services.html
22 https://www.autoblog.com/2011/05/31/women-voice-command-systems/
23 https://www.ncbi.nlm.nih.gov/pubmed/27435949
24 American Roentgen Ray Society (2007), 'Voice Recognition Systems Seem To Make More Errors With Women's Dictation', *ScienceDaily*, 6 May 2007; Rodger, James A. and Pendharkar, Parag C. (2007), 'A field study of database communication issues peculiar to users of a voice activated medical

tracking application', *Decision Support Systems*, 43:1 (1 February 2007), 168–80, https://doi.org/10.1016/j.dss.2006.08.005.

25 American Roentgen Ray Society (2007)

26 http://techland.time.com/2011/06/01/its-not-you-its-it-voice-recognition-doesnt-recognize-women/

27 https://www.ncbi.nlm.nih.gov/pmc/articles/PMC2994697/

28 http://www.aclweb.org/anthology/P08-1044

29 https://www.ncbi.nlm.nih.gov/pmc/articles/PMC2790192/

30 http://www.aclweb.org/anthology/P08-1044

31 http://groups.inf.ed.ac.uk/ami/corpus/; http://www1.icsi.berkeley.edu/Speech/papers/gelbart-ms/numbers/; http://www.voxforge.org/

32 http://www.natcorp.ox.ac.uk/corpus/index.xml?ID=intro

33 http://www.natcorp.ox.ac.uk/docs/URG/BNCdes.html#body.1_div.1_div.5_div.1

34 https://corpus.byu.edu/bnc/5

35 For example: he = 633,413, she = 350,294, himself = 28,696, herself = 15,751

36 He = 3,825,660, she = 2,002,536, himself = 140,087, herself = 70,509

37 Chang, K., Ordonez, V., Wang, T., Yatskar, M. and Zhao, J. (2017), 'Men Also Like Shopping: Reducing Gender Bias Amplification using Corpus-level Constraints', *CoRR*, abs/1707.09457.

38 https://www.eurekalert.org/pub_releases/2015–04/uow-wac040915.php

39 Caliskan, A., Bryson, J. J. and Narayanan, A. (2017), 'Semantics derived automatically from language corpora contain human-like biases', *Science*, 356:6334, 183–6, https://doi.org/10.1126/science.aal4230

40 Bolukbasi, Tolga, Chang, Kai-Wei, Zou, James, Saligrama, Venkatesh and Kalai, Adam (2016), 'Man is to Computer Programmer as Woman is to Homemaker? Debiasing Word Embeddings', 30th Conference on Neural Information Processing Systems (NIPS 2016), Barcelona, http://papers.nips.cc/paper/6228-man-is-to-computer-programmer-as-woman-is-to-homemaker-debiasing-word-embeddings.pdf

41 Chang et al. (2017)

42 https://www.wired.com/story/machines-taught-by-photos-learn-a-sexist-view-of-women?mbid=social_fb

43 https://metode.org/issues/monographs/londa-schiebinger.html

44 https://phys.org/news/2016–09-gender-bias-algorithms.html

45 https://www.theguardian.com/science/2016/sep/01/how-algorithms-rule-our-working-lives

46 https://www.theguardian.com/technology/2018/mar/04/robots-screen-candidates-for-jobs-artificial-intelligence?CMP=twt_gu
47 https://www.techemergence.com/machine-learning-medical-diagnostics-4-current-applications/
48 http://www.bbc.co.uk/news/health-42357257
49 Bolukbasi et al. (2016)

Chapter 9

1 https://www.bloomberg.com/amp/news/articles/2017–09-21/a-smart-breast-pump-mothers-love-it-vcs-don-t
2 Ibid.
3 https://www.newyorker.com/business/currency/why-arent-mothers-worth-anything-to-venture-capitalists/amp
4 Ibid.
5 Ibid.
6 Ibid.
7 https://hbr.org/2017/05/we-recorded-vcs-conversations-and-analyzed-how-differently-they-talk-about-female-entrepreneurs
8 https://www.newyorker.com/business/currency/why-arent-mothers-worth-anything-to-venture-capitalists/amp
9 https://www.bcg.com/publications/2018/why-women-owned-startups-are-better-bet.aspx
10 https://www.bi.edu/research/business-review/articles/2014/03/personality-for-leadership/
11 https://www.bcg.com/publications/2018/how-diverse-leadership-teams-boost-innovation.aspx
12 http://www.bbc.co.uk/news/health-39567240
13 http://blogs.wsj.com/accelerators/2014/08/08/theresia-gouw-no-more-pipeline-excuses/
14 http://science.sciencemag.org/content/355/6323/389
15 http://www.theverge.com/2014/9/25/6844021/apple-promised-an-expansive-health-app-so-why-cant-i-track
16 https://www.theatlantic.com/technology/archive/2014/12/how-self-tracking-apps-exclude-women/383673/; http://www.theverge.com/2014/9/25/6844021/apple-promised-an-expansive-health-app-so-why-cant-i-track; http://www.techtimes.com/articles/16574/20140926/apple-healthkit-period-

tracker.htm; http://nymag.com/thecut/2014/09/new-iphone-grossed-out-by-our-periods.html

17 http://www.telegraph.co.uk/technology/news/8930130/Apple-iPhone-search-Siri-helps-users-find-prostitutes-and-Viagra-but-not-an-abortion.html

18 https://well.blogs.nytimes.com/2016/03/14/hey-siri-can-i-rely-on-you-in-a-crisis-not-always-a-study-finds/

19 https://medium.com/hh-design/the-world-is-designed-for-men-d06640654491#.piekpq2tt

20 https://www.theatlantic.com/technology/archive/2014/12/how-self-tracking-apps-exclude-women/383673/

21 Lupton, Deborah (2015), 'Quantified sex: a critical analysis of sexual and reproductive self-tracking using apps', *Culture, Health & Sexuality*, 17:4

22 Nelson, M. Benjamin, Kaminsky, Leonard A., D. Dickin, Clark and Montoye, Alexander H. K. (2016), 'Validity of Consumer-Based Physical Activity Monitors for Specific Activity Types', *Medicine & Science in Sports & Exercise*, 48:8, 1619–28

23 Murakami, H, Kawakami, R., Nakae, S., Nakata, Y., Ishikawa-Takata, K., Tanaka, S. and Miyachi, M. (2016), 'Accuracy of Wearable Devices for Estimating Total Energy Expenditure: Comparison With Metabolic Chamber and Doubly Labeled Water Method', *JAMA Internal Medicine*, 176:5, 702–3

24 http://genderedinnovations.stanford.edu/case-studies/robots.html#tabs-2

25 Wolfson, Leslie, Whipple, Robert, Derby, Carl A., Amerman, Paula and Nashner, Lewis (1994), Gender Differences in the Balance of Healthy Elderly as Demonstrated by Dynamic Posturography, *Journal of Gerontology*, 49:4, 160–167; Stevens, J. A. and Sogolow, E. D. (2005), 'Gender differences for non-fatal unintentional fall related injuries among older adults', *Injury Prevention*, 11, 115–19

26 Ibid.

27 https://www.ncbi.nlm.nih.gov/pmc/articles/PMC4750302/

28 Chang, Vicky C. and Minh, T. (2015), 'Risk Factors for Falls Among Seniors: Implications of Gender', *American Journal of Epidemiology*, 181:7, 521–31

29 Yin, Hujun et al eds. (2016) *Intelligent Data Engineering and Automated Learning*, Proceedings of the 17th International Conference, Yangzhou China

30 https://www.theatlantic.com/technology/archive/2014/12/how-self-tracking-apps-exclude-women/383673/

31 https://www.afdb.org/en/blogs/investing-in-gender-equality-for-africa%E2%80%99s-transformation/post/technology-women-and-africa-access-use-creation-and-leadership-13999/]

32 https://www.bloomberg.com/news/articles/2016–06-23/artificial-intelligence-has-a-sea-of-dudes-problem

33 http://interactions.acm.org/archive/view/january-february-2014/are-you-sure-your-software-is-gender-neutral

34 http://foreignpolicy.com/2017/01/16/women-vs-the-machine/

35 https://www.bloomberg.com/news/articles/2016–06-23/artificial-intelligence-has-a-sea-of-dudes-problem

36 https://www.ncwit.org/sites/default/files/resources/btn_03232017_web.pdf

37 https://www.ft.com/content/ca324dcc-dcb0-11e6-86ac-f253db7791c6

38 https://www.theverge.com/2016/1/11/10749932/vr-hardware-needs-to-fit-women-too

39 https://mic.com/articles/142579/virtual-reality-has-a-sexual-harassment-problem-what-can-we-do-to-stop-it#.ISQgjAanK

40 https://mic.com/articles/157415/my-first-virtual-reality-groping-sexual-assault-in-vr-harassment-in-tech-jordan-belamire#.5lnAqHFW1

41 http://uploadvr.com/dealing-with-harassment-in-vr/

42 Ibid.

43 https://www.newscientist.com/article/2115648-posture-could-explain-why-women-get-more-vr-sickness-than-men/

44 https://www.newscientist.com/article/dn3628-women-need-widescreen-for-virtual-navigation

45 https://qz.com/192874/is-the-oculus-rift-designed-to-be-sexist/

46 https://www.washingtonpost.com/local/trafficandcommuting/female-dummy-makes-her-mark-on-male-dominated-crash-tests/2012/03/07/gIQANBLjaS_story.html?utm_term=.5ec23738142a

47 'Gendered Innovations: How Gender Analysis Contributes to Research' (2013), report of the Expert Group 'Innovation Through Gender' (chairperson: Londa Schiebinger, rapporteur: Ineke Klinge), Directorate General for Research and Innovation, Luxembourg: Publications Office of the European Union

48 https://crashstats.nhtsa.dot.gov/Api/Public/ViewPublication/811766

49 https://www.washingtonpost.com/local/trafficandcommuting/female-dummy-makes-her-mark-on-male-dominated-crash-tests/2012/03/07/gIQANBLjaS_story.html?utm_term=.5ec23738142a

50 http://genderedinnovations.stanford.edu/case-studies/crash.html#tabs-2

51 Ibid.

52 https://www.washingtonpost.com/local/trafficandcommuting/female-dummy-makes-her-mark-on-male-dominated-crash-tests/2012/03/07/gIQANBLjaS_story.html?utm_term=.5ec23738142a
53 Linder, Astrid and Svedberg, Wanna (2018), 'Occupant Safety Assessment in European Regulatory Tests: Review of Occupant Models, Gaps and Suggestion for Bridging Any Gaps', conference paper, 'Road Safety on Five Continents' South Korea, May 2018
54 http://sciencenordic.com/gender-equality-crash-test-dummies-too
55 Linder and Svedberg (2018)
56 United States Government Publishing Office, 'U.S. Code of Federal Regulations. 2011. 49 CFR U, -2RE Side Impact Crash Test Dummy, 50th Percentile Adult Male'. http://www.gpo.gov/fdsys/granule/CFR-2011-title49-vol7/CFR- 2011-title49-vol7-part572-subpartU
57 Linder and Svedberg (2018)
58 http://genderedinnovations.stanford.edu/case-studies/crash.html#tabs-2
59 http://media.leidenuniv.nl/legacy/leru-paper-gendered-research-and-innovation.pdf; Londa Schiebinger and Martina Schraudner (2011), 'Interdisciplinary Approaches to Achieving Gendered Innovations in Science, Medicine, and Engineering', *Interdisciplinary Science Reviews*, 36:2 (June 2011), 154–67
60 http://genderedinnovations.stanford.edu/case-studies/crash.html#tabs-2
61 'Gendered Innovations: How Gender Analysis Contributes to Research' (2013)
62 http://genderedinnovations.stanford.edu/case-studies/crash.html#tabs-2
63 https://www.washingtonpost.com/local/trafficandcommuting/female-dummy-makes-her-mark-on-male-dominated-crash-tests/2012/03/07/gIQANBLjaS_story.html?utm_term=.5ec23738142a
64 'Gendered Innovations: How Gender Analysis Contributes to Research' (2013)
65 http://content.tfl.gov.uk/travel-in-london-understanding-our-diverse-communities.pdf; http://www.wnyc.org/story/283137-census-data-show-public-transit-gender-gap/
66 https://eur-lex.europa.eu/resource.html?uri=cellar:41f89a28-1fc6-4c92-b1c8-03327d1b1ecc.0007.02/DOC_1&format=PDF

Chapter 10

1 Marts, Sherry A. and Keitt, Sarah (2004), 'Principles of Sex-based Differences in Physiology: Foreword: a historical overview of advocacy for research in sex based biology', *Advances in Molecular and Cell Biology*, 34, 1–333

2 A Medline search shows that the male norm remains, as practice guidelines and research examples are still often expressed in terms of the typical 70 kg man: Marts and Keitt (2004)

3 Pages 17–18

4 Plataforma SINC (2008), 'Medical Textbooks Use White, Heterosexual Men As A 'Universal Model', *ScienceDaily*, <www.sciencedaily.com/releases/2008/10/081015132108.htm>.

5 Dijkstra, A. F, Verdonk, P. and Lagro-Janssen, A. L. M. (2008), 'Gender bias in medical textbooks: examples from coronary heart disease, depression, alcohol abuse and pharmacology', *Medical Education*, 42:10, 1021–8

6 http://www.marieclaire.com/health-fitness/a26741/doctors-treat-women-like-men/; https://www.health.harvard.edu/heart-health/gender-matters-heart-disease-risk-in-women

7 Dijkstra et al. (2008)

8 Henrich, Janet B. and Viscoli, Catherine M. (2006), 'What Do Medical Schools Teach about Women's Health and Gender Differences?' *Academic Medicine*, 81:5

9 Song, Michael M. Jones, Betsy G. and Casanova, Robert A. (2016), 'Auditing sex- and gender-based medicine (SGBM) content in medical school curriculum: a student scholar model', *Biology of Sex Differences*, 7:Suppl 1, 40

10 Marts and Keitt (2004)

11 Karp, Natasha A. et al (2017), 'Prevalence of sexual dimorphism in mammalian phenotypic traits', *Nature Communications*, 8:15475

12 Martha L. Blair (2007), 'Sex-based differences in physiology: what should we teach in the medical curriculum?', *Advanced Physiological Education*, 31, 23–5

13 Ibid.

14 https://www.ncbi.nlm.nih.gov/pmc/articles/PMC4800017/ Jan 2016

15 https://theconversation.com/man-flu-is-real-but-women-get-more-autoimmune-diseases-and-allergies-77248

16 https://www.washingtonpost.com/national/health-science/why-do-autoimmune-diseases-affect-women-more-often-than-men/2016/10/17/3e224db2-8429-11e6-ac72-a29979381495_story.html?utm_term=.acef157fc395

17 http://www.nature.com/news/infections-reveal-inequality-between-the-sexes-1.20131?WT.mc_id=TWT_NatureNews

18 Ibid.

19 https://www.ncbi.nlm.nih.gov/pmc/articles/PMC4157517/

20 Ibid.

21 http://docs.autismresearchcentre.com/papers/2010_Schwartz_SexSpecific_MolAut.pdf

22 Clayton, Janine Austin (2015), 'Studying both sexes: a guiding principle for biomedicine', http://www.fasebj.org/content/early/2015/10/28/fj.15-279554.full.pdf+html

23 Ibid.

24 Ibid.

25 https://theconversation.com/not-just-about-sex-throughout-our-bodies-thousands-of-genes-act-differently-in-men-and-women-86613

26 Holdcroft, Anita, Snidvongs, Saowarat and Berkley, Karen J. (2011), 'Incorporating Gender and Sex Dimensions in Medical Research', *Interdisciplinary Science Reviews*, 36:2, 180–92

27 'Gender and Health Knowledge Agenda', May 2015, ZonMw, Netherlands, http://www.genderportal.eu/sites/default/files/resource_pool/Gender%20%26%20Health%20Knowledge%20Agenda_0.pdf

28 Pollitzer, Elizabeth (2013), 'Cell sex matters', *Nature*, 500, 23–24

29 Londa Schiebinger (2014), 'Gendered innovations: harnessing the creative power of sex and gender analysis to discover new ideas and develop new technologies', *Triple Helix*, 1:9

30 Cristiana Vitale et al. (2017), 'Under-representation of elderly and women in clinical trials', *International Journal of Cardiology*, 232, 216–21

31 The Henry J. Kaiser Family Foundation (2014), 'Women and HIV/AIDS in the United States'; women also experience different clinical symptoms and complications due to HIV disease

32 http://www.who.int/gender/hiv_aids/hivaids1103.pdf

33 Curno, Mirjam J. et al. (2016), 'A Systematic Review of the Inclusion (or Exclusion) of Women in HIV Research: From Clinical Studies of Antiretrovirals and Vaccines to Cure Strategies', *Journal of Acquired Immune Deficiency Syndrome*, 1:71(2) (February 2016),. 181–8

34 http://www.wpro.who.int/topics/gender_issues/Takingsexandgenderintoaccount.pdf

35 Ibid.

36 Hughes, Robert N. (2007), 'Sex does matter: comments on the prevalence of male-only investigations of drug effects on rodent behaviour', *Behavioural Pharmacology*, 18:7, 583–9

37 http://helix.northwestern.edu/article/thalidomide-tragedy-lessons-drug-safety-and-regulation

38 https://www.smh.com.au/national/the-50-year-global-cover-up-20120725-22r5c.html

39 http://broughttolife.sciencemuseum.org.uk/broughttolife/themes/controversies/thalidomide

40 Marts and Keitt (2004)
41 http://foreignpolicy.com/2014/08/20/why-are-so-many-women-dying-from-ebola/
42 R. D. Fields (2014), 'Vive la différence requiring medical researchers to test males and females in every experiment sounds reasonable, but it is a bad idea', *Scientific American*, 311, 14
43 Richardson, S. S., Reiches, M., Shattuck-Heidorn, H., LaBonte, M. L. and Consoli, T. (2015), 'Opinion: focus on preclinical sex differences will not address women's and men's health disparities', *Proceedings of the National Academy of Science*, 112, 13419–20
44 Holdcroft, Anita (2007) 'Gender bias in research: how does it affect evidence based medicine?', *Journal of the Royal Society of Medicine*, 100
45 Ibarra, Manuel, Vázquez, Marta and Fagiolino, Pietro (2017), 'Sex Effect on Average Bioequivalence', *Clinical Therapeutics*, 39:1, 23–33
46 Mergaert, Lut and Lombardo, Emanuela (2014), 'Resistance to implementing gender mainstreaming in EU research policy', in Weiner, Elaine and MacRae, Heather (eds.), 'The persistent invisibility of gender in EU policy', European Integration online Papers (EIoP), special issue 1, Vol. 18, Article 5, 1–21
47 Ibid.
48 Ibid.
49 Hughes (2007)
50 Pinnow, Ellen, Herz, Naomi, Loyo-Berrios, Nilsa and Tarver, Michelle (2014), 'Enrollment and Monitoring of Women in Post-Approval Studies for Medical Devices Mandated by the Food and Drug Adminstration', *Journal of Women's Health*, 23:3 (March 2014), 218–23
51 http://www.sciencedirect.com/science/article/pii/S0002870310000864
52 Labots, G., Jones, A., Visser, S. J. de, Rissmann, R. and Burggraaf, J. (2018), 'Gender differences in clinical registration trials: is there a real problem?', *British Journal of Clinical Pharmacology*
53 McGregor, Alyson J. (2017), 'The Effects of Sex and Gender on Pharmacologic Toxicity: Implications for Clinical Therapy', *Clinical Therapeutics*, 39:1
54 Ibid.
55 Ibid.
56 Ibid.
57 Bruinvels, G. et al. (2016), 'Sport, exercise and the menstrual cycle: where is the research?', *British Journal of Sports Medicine*, 51:6, 487–488
58 Zopf, Y. et al. (2008), 'Women encounter ADRs more often than do men', *European Journal of Clinical Pharmacology*, 64:999

59 https://www.ncbi.nlm.nih.gov/pmc/articles/PMC198535/

60 Soldin, Offie P., Chung, Sarah H. and Mattison, Donald R. (2011), 'Sex Differences in Drug Disposition', *Journal of Biomedicine and Biotechnology*, 2011:187103; Anderson, Gail D. (2005), 'Sex And Racial Differences In Pharmacological Response: Where Is The Evidence? Pharmacogenetics, Pharmacokinetics, and Pharmacodynamics', *Journal of Women's Health*, 14:1, http://online.liebertpub.com.libproxy.ucl.ac.uk/doi/pdf/10.1089/jwh.2005.14.19

61 Anderson (2005)

62 Hughes (2007)

63 Yoon, Dustin Y. et al. (2014), 'Sex bias exists in basic science and translational surgical research', *Surgery*, 156:3, 508–16

64 https://thinkprogress.org/scientists-avoid-studying-womens-bodies-because-they-get-periods-3fe9d6c39268/

65 Yoon et al. (2014)

66 Karp (2017)

67 Hughes (2007)

68 Yoon et al. (2014)

69 Ibid.

70 Ibid.

71 Ortona, Elena, Delunardo, Federica, Baggio, Giovannella and Malorni, Walter (2016), 'A sex and gender perspective in medicine: A new Mandatory Challenge For Human Health', *Ann Ist Super Sanità*, 52:2, 146–8

72 J. Peretz et al. (2016), 'Estrogenic compounds reduce influenza A virus in primary human nasal epithelial cells derived from female, but not male, donors', *American Journal of Physiology*, 310:5, 415–425

73 http://protomag.com/articles/pain-women-pain-men

74 https://www.newscientist.com/article/dn28064-female-viagra-has-been-approved-heres-what-you-need-to-know/

75 Anderson (2005); Whitley, Heather P. and Lindsey, Wesley (2009), 'Sex-Based Differences in Drug Activity', *American Family Physician*, 80:11 (December 2009), 1254–8

76 https://www.accessdata.fda.gov/drugsatfda_docs/label/2015/022526 REMS.pdf

77 https://www.ncbi.nlm.nih.gov/pmc/articles/PMC4800017/ https://www.ncbi.nlm.nih.gov/pubmed/20799923; Howard, Louise M., Ehrlich, Anna M., Gamlen, Freya and Oram, Sian (2017), 'Gender-neutral mental health research is sex and gender biased', *Lancet Psychiatry*, 4:1, 9–11

78 Marts and Keitt (2004)

79 Parekh, A., Sanhai, W., Marts, S. and Uhl, K. (2007), 'Advancing women's health via FDA Critical Path Initiative', *Drug Discovery Today: Technologies*, 4:2

80 http://www.nature.com/news/infections-reveal-inequality-between-the-sexes-1.20131?WT.mc_id=TWT_NatureNews

81 Yoon et al. (2014)

82 http://genderedinnovations.stanford.edu/case-studies/colon.html#tabs-2

83 Devries, Michaela C. (2016), 'Sex-based differences in endurance exercise muscle metabolism: impact on exercise and nutritional strategies to optimize health and performance in women', *Experimental Physiology*, 101:2, 243–9

84 Schiebinger (2014)

85 Zusterzeel, R. et al. (2014), 'Cardiac Resynchronization Therapy in Women: US Food and Drug Administration Meta-analysis of Patient-Level Data', *JAMA Internal Medicine*, 174:8, 1340–8

86 Woodruff, Teresa K. (2014), 'Sex, equity, and science', *PNAS*, 111:14, 5,063–4

87 Nowak, Bernd et al. (2010), 'Do gender differences exist in pacemaker implantation?–results of an obligatory external quality control program', *Europace*, 12, 210–15

88 http://www.smithsonianmag.com/innovation/the-worlds-first-true-artificial-heart-now-beats-inside-a-75-year-old-patient-180948280/?no-ist

89 http://www.syncardia.com/medical-professionals/two-sizes-70cc-50cc.html

90 Sardeli, Amanda Veiga and Chacon-Mikahil, Mara Patricia T. (2016), 'Exercise-Induced Increase as a Risk Factor for Central Arterial Stiffness', Journal of Archives in Military Medicine; http://circ.ahajournals.org/content/110/18/2858; http://www.medscape.com/viewarticle/728571; https://www.ncbi.nlm.nih.gov/pubmed/22267567

91 Collier, Scott R. (2008), 'Sex Differences in the Effects of Aerobic and Anaerobic Exercise on Blood Pressure and Arterial Stiffness', *Gender Medicine*, 5:2

92 Ibid.

93 Devries (2016)

94 Tarnopolsky, M. A. (2008), 'Sex Differences in Exercise Metabolism and the Role of 17-Beta Estradiol', *Medicine and Science in Sports and Exercise*, 40:4, 648–54

95 Dick, R. W. (2009), 'Is there a gender difference in concussion incidence and outcomes?', *British Journal of Sports Medicine*, 43, Suppl. I, i46–i50, DOI:10.1136/bjsm.2009.058172

96 https://thinkprogress.org/scientists-avoid-studying-womens-bodies-because-they-get-periods-3fe9d6c39268/)

97 Hunter, Sandra K. (2016), 'Sex differences in fatigability of dynamic contractions', *Experimental Physiology*, 101:2, 250–5

98 Jutte, Lisa S., Hawkins, Jeremy, Miller, Kevin C., Long, Blaine C. and Knight, Kenneth L. (2012), 'Skinfold Thickness at 8 Common Cryotherapy Sites in Various Athletic Populations', *Journal of Athletic Training*, 47:2, 170–7

99 Costello, Joseph T., Bieuzen, Francois and Bleakley, Chris M. (2014), 'Where are all the female participants in Sports and Exercise Medicine research?', *European Journal of Sport Science*, 14:8, 847–51; https://www.sciencenews.org/blog/scicurious/women-sports-are-often-underrepresented-science

100 Faulkner, S. H., Jackson, S., Fatania, G. and Leicht, C. A. (2017), 'The effect of passive heating on heat shock protein 70 and interleukin-6: A possible treatment tool for metabolic diseases?', *Temperature*, 4, 1–13

101 https://theconversation.com/a-hot-bath-has-benefits-similar-to-exercise-74600; http://www.huffingtonpost.com/entry/hot-bath-may-have-similar-benefits-as-exercise_us_58d90aa8e4b03692bea7a930

102 'Gender and Health Knowledge Agenda', May 2015 (ZonMw, Netherlands); http://www.health.harvard.edu/heart-health/gender-matters-heart-disease-risk-in-women; Dallongeville, J. et al. (2010), 'Gender differences in the implementation of cardiovascular prevention measures after an acute coronary event', *Heart*, 96, 1744–9

103 'Gender and Health Knowledge Agenda', May 2015 (ZonMw, Netherlands)

104 https://theconversation.com/medicines-gender-revolution-how-women-stopped-being-treated-as-small-men-77171

105 https://orwh.od.nih.gov/clinical/women-and-minorities/

106 https://orwh.od.nih.gov/sites/orwh/files/docs/NOT-OD-15–102_Guidance.pdf

107 Yoon et al. (2014)

108 Rees, Teresa (2011), 'The Gendered Construction of Scientific Excellence', *Interdisciplinary Science Reviews*, 36:2, 133–45

109 Howard, Ehrlich, Gamlen and Oram (2017)

110 Holdcroft (2007)

111 Ibid.

112 Marts and Keitt (2004)

113 http://www.nature.com/news/infections-reveal-inequality-between-the-sexes-1.20131?WT.mc_id=TWT_NatureNews

114 Ortona, Delunardo, Baggio and Malorni (2016)

115 http://www.goretro.com/2014/08/mothers-little-helper-vintage-drug-ads.html

116 https://www.ncbi.nlm.nih.gov/pmc/articles/PMC198535/

117 Ibid.

118 'Gender and Health Knowledge Agenda', May 2015 (ZonMw, Netherlands)

119 https://www.ncbi.nlm.nih.gov/pmc/articles/PMC198535/

120 Ibid.

121 Ibid.

122 http://www.ajmc.com/newsroom/women-taking-stains-faced-increased-diabetes-risk

123 http://www.health.harvard.edu/heart-health/gender-matters-heart-disease-risk-in-women

124 Pollitzer (2013)

125 https://www.ncbi.nlm.nih.gov/pmc/articles/PMC4800017/ Jan 2016

126 Whitley and Lindsey (2009)

127 https://www.washingtonpost.com/news/wonk/wp/2014/06/07/bad-medicine-the-awful-drug-reactions-americans-report/?utm_term=.1a7067d-40dce

128 Tharpe, N. (2011), 'Adverse Drug Reactions in Women's Health Care', *Journal of Midwifery & Women's Health*, 56, 205–13

129 https://www.washingtonpost.com/news/wonk/wp/2014/06/07/bad-medicine-the-awful-drug-reactions-americans-report/?utm_term=.1a7067d-40dce

130 Marts and Keitt (2004)

131 Carey, Jennifer L. et al. (2017), 'Drugs and Medical Devices: Adverse Events and the Impact on Women's Health', *Clinical Therapeutics*, 39:1

132 Yoon et al. (2014)

133 'Gender and Health Knowledge Agenda', May 2015 (ZonMw, Netherlands)

134 https://www.ncbi.nlm.nih.gov/pmc/articles/PMC198535/

135 https://www.hindawi.com/journals/bmri/2011/187103/

136 Ibid.

137 Anderson (2005)

138 Wang, Lishi et al. (2017), 'Sex Differences in Hazard Ratio During Drug Treatment of Non-small-cell Lung Cancer in Major Clinical Trials: A Focused Data Review and Meta-analysis', *Clinical Therapeutics*, 39:1

139 Ibarra, Vázquez and Fagiolino (2017)

140 Whitley and Lindsey (2009)

141 Ibid.

142 Ibid.

Chapter 11

1 https://www.georgeinstitute.org/media-releases/disadvantaged-women-at-greater-risk-of-heart-disease-than-men-0

2 https://www.ncbi.nlm.nih.gov/pmc/articles/PMC4800017/ Jan 2016; http://circ.ahajournals.org/content/133/9/916?sid=beb5f268-4205-4e62-be8f-3caec4c4d9b7

3 http://heart.bmj.com/content/102/14/1142

4 http://circ.ahajournals.org/content/133/9/916?sid=beb5f268-4205-4e62-be8f-3caec4c4d9b7 2016

5 Ridker, Paul M. et al. (2005), 'A Randomized Trial of Low-Dose Aspirin in the Primary Prevention of Cardiovascular Disease in Women', *New England Journal of Medicine*, 352, 1293–304

6 Johannes, A. N. et al. (2011), 'Aspirin for primary prevention of vascular events in women: individualized prediction of treatment effects', *European Heart Journal*, 32:23, 2962–9

7 Kruijsdijk, R. C. M. van et al. (2015), 'Individualised prediction of alternate-day aspirin treatment effects on the combined risk of cancer, cardiovascular disease and gastrointestinal bleeding in healthy women', *Heart*, 101, 369–76

8 Wu, J. et al. (2016), 'Impact of initial hospital diagnosis on mortality for acute myocardial infarction: A national cohort study', *European Heart Journal*, 7:2

9 https://www.nytimes.com/2014/09/28/opinion/sunday/womens-atypical-heart-attacks.html?_r=0

10 http://heart.bmj.com/content/102/14/1142

11 Ibid.

12 Yoon et al. (2014)

13 http://circ.ahajournals.org/content/133/9/916?sid=beb5f268-4205-4e62-be8f-3caec4c4d9b7

14 https://www.england.nhs.uk/wp-content/uploads/2013/06/a09-cardi-prim-percutaneous.pdf

15 https://www.hqip.org.uk/wp-content/uploads/2018/02/national-audit-of-percutaneous-coronary-intervention-annual-public-report.pdf

16 https://www.sciencedaily.com/releases/2016/03/160304092233.htm 2016 march

17 http://heart.bmj.com/content/102/14/1142 Published online 24 June 2016.

18 https://www.sciencedaily.com/releases/2016/03/160304092233.htm 2016 march

19 Motiwala, Shweta R., Sarma, Amy, Januzzi, James L. and O'Donoghue, Michelle L. (2014), 'Biomarkers in ACS and Heart Failure: Should Men and Women Be Interpreted Differently?', *Clinical Chemistry*, 60:1

20 'Gender and Health Knowledge Agenda', May 2015 (ZonMw, Netherlands)
21 http://media.leidenuniv.nl/legacy/leru-paper-gendered-research-and-innovation.pdf
22 Ibid.
23 'Gender and Health Knowledge Agenda', May 2015 (ZonMw, Netherlands)
24 Schiebinger, Londa (2014), 'Gendered innovations: harnessing the creative power of sex and gender analysis to discover new ideas and develop new technologies', *Triple Helix*, 1:9
25 Dijkstra, A. F, Verdonk, P. and Lagro-Janssen, A. L. M. (2008), 'Gender bias in medical textbooks: examples from coronary heart disease, depression, alcohol abuse and pharmacology', *Medical Education*, 42:10, 1021–8; 'Gender and Health Knowledge Agenda', May 2015 (ZonMw, Netherlands); https://link.springer.com/article/10.1007/s10459-008-9100-z; Holdcroft (2007)
26 Sakalihasan, N., Limet, R. and Defawe, O. D. (2005), 'Abdominal aortic aneurysm', *Lancet*, 365, 1577–89
27 'Gender and Health Knowledge Agenda', May 2015 (ZonMw, Netherlands)
28 http://genderedinnovations.stanford.edu/case-studies/colon.html#tabs-2
29 Ibid.
30 Ibid.
31 http://www.wpro.who.int/topics/gender_issues/Takingsexandgenderintoaccount.pdf
32 United Nations Development Programme (2015), *Discussion Paper: Gender and Tuberculosis*
33 ACTION (Advocacy to Control TB Internationally), 'Women and Tuberculosis: Taking a Look at a Neglected Issue', ACTION, Washington DC, 2010
34 Ibid.
35 United Nations Development Programme (2015), *Discussion Paper: Gender and Tuberculosis*
36 ACTION (Advocacy to Control TB Internationally), 'Women and Tuberculosis: Taking a Look at a Neglected Issue', ACTION, Washington DC, 2010.
37 Ibid.; United Nations Development Programme (2015), *Discussion Paper: Gender and Tuberculosis*
38 ACTION (Advocacy to Control TB Internationally), 'Women and Tuberculosis: Taking a Look at a Neglected Issue', ACTION, Washington DC, 2010; United Nations Development Programme (2015), *Discussion Paper: Gender and Tuberculosis*
39 http://www.wpro.who.int/topics/gender_issues/Takingsexandgenderintoaccount.pdf

40 ACTION (Advocacy to Control TB Internationally), 'Women and Tuberculosis: Taking a Look at a Neglected Issue', ACTION, Washington DC, 2010
41 United Nations Development Programme (2015), *Discussion Paper: Gender and Tuberculosis*
42 ACTION (Advocacy to Control TB Internationally), 'Women and Tuberculosis: Taking a Look at a Neglected Issue', ACTION, Washington DC, 2010.
43 Schiebinger (2014)
44 http://genderedinnovations.stanford.edu/case-studies/hiv.html#tabs-2
45 https://www.scientificamerican.com/article/autism-it-s-different-in-girls/
46 https://www.um.edu.mt/library/oar/handle/123456789/15597
47 https://www.scientificamerican.com/article/autism-it-s-different-in-girls/
48 https://www.um.edu.mt/library/oar/handle/123456789/15597
49 https://www.scientificamerican.com/article/autism-it-s-different-in-girls/
50 https://www.theguardian.com/society/2016/oct/21/m-in-the-middle-girls-autism-publish-novel-limpsfield-grange
51 https://www.gov.uk/government/consultations/adult-autism-strategy-guidance-update
52 https://www.theguardian.com/society/2016/oct/21/m-in-the-middle-girls-autism-publish-novel-limpsfield-grange
53 https://www.theatlantic.com/health/archive/2013/04/adhd-is-different-for-women/381158/?utm_source=quartzfb
54 Hoffman, Diane E. and Tarzian, Anita J. (2001), 'The Girl Who Cried Pain: A Bias Against Women in the Treatment of Pain', *Journal of Law, Medicine & Ethics*, 29, 13–27
55 http://thinkprogress.org/health/2015/05/11/3654568/gender-roles-women-health/
56 https://www.theguardian.com/society/2017/sep/06/listen-to-women-uk-doctors-issued-with-first-guidance-on-endometriosis
57 https://www.endofound.org/endometriosis
58 https://www.theguardian.com/society/2015/sep/28/endometriosis-hidden-suffering-millions-women
59 https://www.theguardian.com/society/2017/sep/06/listen-to-women-uk-doctors-issued-with-first-guidance-on-endometriosis
60 http://www.independent.co.uk/news/science/stephen-hawking-says-women-are-the-most-intriguing-mystery-in-reddit-ama-a6687246.html
61 https://www.birdvilleschools.net/cms/lib/TX01000797/Centricity/Domain/1013/AP%20Psychology/Femininity.pdf
62 Showalter, Elaine (1985) *The Female Malady: Women, Madness and English Culture 1830–1980*, London 1987

63 https://www.health.harvard.edu/blog/astounding-increase-in-antidepressant-use-by-americans-201110203624
64 http://pb.rcpsych.org/content/pbrcpsych/early/2017/01/06/pb.bp.116.054270.full.pdf
65 https://academic.oup.com/painmedicine/article/10/2/289/article
66 Hoffman and Tarzian (2001)
67 Fillingim, R. B., King, C. D., Ribeiro-Dasilva, M. C., Rahim-Williams, B. and Riley, J. L. (2009), 'Sex, Gender, and Pain: A Review of Recent Clinical and Experimental Findings', *Journal of Pain: Official Journal of the American Pain Society*, 10:5, 447–85
68 https://www.med.unc.edu/ibs/files/educational-gi-handouts/IBS%20in%20Women.pdf
69 https://www.npr.org/sections/health-shots/2012/04/16/150525391/why-women-suffer-more-migraines-than-men
70 https://migraine.com/migraine-statistics/
71 http://www.independent.co.uk/life-style/health-and-families/health-news/will-this-hurt-doctor-much-more-if-you-are-a-woman-907220.html
72 Greenspan, Joel D. et al. (2007), 'Studying sex and gender differences in pain and analgesia: A consensus report', *Pain*, 132, S26–S45
73 Hoffmann and Tarzian (2001)
74 Clayton, Janine Austin (2016), 'Studying both sexes: a guiding principle for biomedicine', *The FASEB Journal*, 30:2, 519–524
75 http://www.independent.co.uk/life-style/health-and-families/health-news/will-this-hurt-doctor-much-more-if-you-are-a-woman-907220.html
76 http://www.npr.org/templates/story/story.php?storyId=18106275
77 Ibid.
78 https://www.ncbi.nlm.nih.gov/books/NBK92516/
79 http://www.gendermedicine.com/1st/images/Oral02.pdf
80 Kindig, David A. and Cheng, Erika R. (2013), 'Even As Mortality Fell In Most US Counties, Female Mortality Nonetheless Rose In 42.8 Percent Of Counties From 1992 To 2006', *Health Affairs*, 32:3, 451–8
81 'Gender and Health Knowledge Agenda', May 2015 (ZonMw, Netherlands)
82 http://ajph.aphapublications.org/doi/10.2105/AJPH.2016.303089
83 Ibid.
84 https://www.newscientist.com/article/2081497-women-live-longer-than-men-but-suffer-more-years-of-poor-health/
85 https://link.springer.com/article/10.1007%2Fs10433-008-0082-8

86 http://www.demographic-research.org/volumes/vol20/19/20–19.pdf
87 http://www.euro.who.int/__data/assets/pdf_file/0006/318147/EWHR16_interactive2.pdf?ua=1
88 https://www.researchgate.net/blog/post/why-do-we-still-not-know-what-causes-pms
89 https://www.nhs.uk/conditions/erection-problems-erectile-dysfunction/treatment/
90 https://www.health.harvard.edu/womens-health/treating-premenstrual-dysphoric-disorder
91 https://www.researchgate.net/blog/post/why-do-we-still-not-know-what-causes-pms
92 http://grantome.com/grant/NIH/R03-TW007438-02
93 https://qz.com/611774/period-pain-can-be-as-bad-as-a-heart-attack-so-why-arent-we-researching-how-to-treat-it/ 2016
94 Ibid.
95 http://grantome.com/grant/NIH/R03-TW007438-02
96 Dmitrovic, R., Kunselman, A. R. and Legro, R. S. (2013), 'Sildenafil citrate in the treatment of pain in primary dysmenorrhea: a randomized controlled trial', *Human Reproduction*, 28:11, 2958–65
97 http://edition.cnn.com/2013/03/27/health/viagra-anniversary-timeline/index.html
98 http://www.clevelandclinicmeded.com/medicalpubs/diseasemanagement/endocrinology/erectile-dysfunction/
99 http://edition.cnn.com/2013/03/27/health/viagra-anniversary-timeline/index.html
100 http://www.telegraph.co.uk/women/life/period-pain-can-feel-bad-heart-attack-ignored/
101 http://www.who.int/mediacentre/factsheets/fs348/en/
102 https://www.pri.org/stories/2017–05-05/how-trumps-latest-budget-impacts-women-and-girls-classrooms-cops
103 https://livestream.com/refinerytv/physiology2016/videos/131487028
104 https://www.propublica.org/article/nothing-protects-black-women-from-dying-in-pregnancy-and-childbirth
105 https://edition.cnn.com/2018/02/20/opinions/protect-mother-pregnancy-williams-opinion/index.html
106 https://www.ncbi.nlm.nih.gov/pubmed/26444126

Chapter 12

1 https://www.thetimes.co.uk/article/review-the-growth-delusion-the-wealth-and-wellbeing-of-nations-by-david-pilling-b322223kc
2 https://www.chathamhouse.org/expert/comment/g20-must-push-more-inclusive-gdp
3 https://www.theguardian.com/uk-news/2016/nov/10/doing-the-chores-valued-at-1tn-a-year-in-the-uk
4 http://databank.worldbank.org/data/download/GDP.pdf
5 http://www.oecd.org/dev/development-gender/Unpaid_care_work.pdf
6 http://progress.unwomen.org/en/2015/pdf/UNW_progressreport.pdf
7 https://www.theatlantic.com/business/archive/2016/03/unpaid-caregivers/474894/
8 https://www.theatlantic.com/business/archive/2016/03/unpaid-caregivers/474894/
9 http://progress.unwomen.org/en/2015/pdf/UNW_progressreport.pdf
10 http://www.pwc.com.au/australia-in-transition/publications/understanding-the-unpaid-economy-mar17.pdf
11 http://hdr.undp.org/sites/default/files/folbre_hdr_2015_final_0.pdf
12 Ibid.
13 Sánchez de Madariaga, Inés, 'Mobility of Care: Introducing New Concepts in Urban Transport', in Marion Roberts and Inés Sánchez de Madariaga (eds.) (2013), *Fair Shared Cities: The Impact of Gender Planning in Europe*, Farnham
14 http://hdr.undp.org/sites/default/files/folbre_hdr_2015_final_0.pdf
15 http://progress.unwomen.org/en/2015/pdf/UNW_progressreport.pdf
16 In 2011 the OECD's annual Social Indicators report included a chapter on unpaid work – but no report since has done the same. http://www.oecd-ilibrary.org/docserver/download/8111041e.pdf?expires=1500914228&id=id&accname=guest&checksum=CD8E8A5F41FA84BE66F2291FF893E9F0
17 https://theconversation.com/gender-neutral-policies-are-a-myth-why-we-need-a-womens-budget-55231
18 Himmelweit, Susan (2002), 'Making Visible the Hidden Economy: The Case for Gender-Impact Analysis of Economic Policy', *Feminist Economics*, 8:1, 49–70, http://dx.doi.org/10.1080/13545700110104864
19 https://www.unison.org.uk/content/uploads/2014/06/On-line-Catalogue224222.pdf
20 http://wbg.org.uk/wp-content/uploads/2017/03/WBG_briefing_Social-Care_Budget-2017_final_JDH_SH_EN_20Mar.pdf
21 Ibid.

22 Ibid.

23 https://www.theguardian.com/lifeandstyle/2012/mar/18/public-sector-cuts-hit-prudent-houseife

24 https://www.unison.org.uk/content/uploads/2014/06/On-line-Catalogue224222.pdf

25 http://wbg.org.uk/wp-content/uploads/2017/03/WBG_briefing_Soc-Security_pre_Budget.pdf

26 https://www.theguardian.com/commentisfree/2017/may/01/conservatives-universal-credit-hard-work

27 https://wbg.org.uk/news/low-income-women-lose-2000-tax-benefit-changes/

28 http://progress.unwomen.org/en/2015/pdf/UNW_progressreport.pdf; http://wbg.org.uk/wp-content/uploads/2016/12/WBG_Budget2017_Fullresponse-1.pdf

29 http://wbg.org.uk/wp-content/uploads/2016/12/Budget_pressrelease_9Mar17.pdf

30 https://www.legislation.gov.uk/ukpga/2010/15/section/149

31 https://www.theguardian.com/commentisfree/2017/may/01/conservatives-universal-credit-hard-work

32 Barsh, Joanna and Yee, Lareina (2011), 'Unlocking the full potential of women in the U.S. Economy', McKinsey

33 http://reports.weforum.org/global-gender-gap-report-2015/the-case-for-gender-equality/

34 http://ec.europa.eu/eurostat/statistics-explained/index.php/Gender_statistics#Labour_market

35 https://data.worldbank.org/indicator/SL.TLF.CACT.FE.ZS?locations=US, accessed 13 March 2018

36 Ibid.; https://data.worldbank.org/indicator/SL.TLF.CACT.MA.ZS, accessed 13 March 2018

37 http://reports.weforum.org/global-gender-gap-report-2015/the-case-for-gender-equality/

38 McKinsey (2015), 'The Power of Parity: how advancing women's equality can add $12 trillion to global growth'

39 Ibid.

40 Ibid.

41 http://progress.unwomen.org/en/2015/pdf/UNW_progressreport.pdf

42 Himmelweit (2002)

43 http://cep.lse.ac.uk/pubs/download/dp1464.pdf

44 https://www.researchgate.net/publication/269288731_Business_training_plus_for_female_entrepreneurship_Short_and_medium-term_experimental_evidence_from_Peru
45 http://www.salute.gov.it/imgs/C_17_pagineAree_431_listaFile_itemName_1_file.pdf
46 http://progress.unwomen.org/en/2015/pdf/UNW_progressreport.pdf
47 http://www.who.int/mediacentre/news/releases/2014/lancet-ageing-series/en/
48 Ibid.
49 https://www.kingsfund.org.uk/projects/time-think-differently/trends-disease-and-disability-long-term-conditions-multi-morbidity
50 http://ec.europa.eu/eurostat/tgm/table.do?tab=table&language=en&pcode=tps00001&tableSelection=1&footnotes=yes&labeling=labels&plugin=1
51 http://www.salute.gov.it/imgs/C_17_pagineAree_431_listaFile_itemName_1_file.pdf
52 Ibid.
53 http://caringeconomy.org/wp-content/uploads/2015/08/care-crisis-means-big-trouble.pdf
54 Ibid.
55 http://www.slate.com/blogs/xx_factor/2017/06/20/the_gop_s_plan_to_slash_medicaid_will_shift_a_costly_burden_onto_women_who.html
56 http://progress.unwomen.org/en/2015/pdf/UNW_progressreport.pdf
57 https://www.alzheimersresearchuk.org/wp-content/uploads/2015/03/Women-and-Dementia-A-Marginalised-Majority1.pdf
58 Ibid.
59 http://www.mckinsey.com/global-themes/gender-equality/the-power-of-parity-advancing-womens-equality-in-the-united-kingdom
60 http://www.nytimes.com/2010/06/10/world/europe/10iht-sweden.html
61 https://iwpr.org/publications/impact-equal-pay-poverty-economy/
62 http://wbg.org.uk/wp-content/uploads/2016/11/De_Henau_Perrons_WBG_CareEconomy_ITUC_briefing_final.pdf
63 http://wbg.org.uk/wp-content/uploads/2016/11/De_Henau_WBG_childcare_briefing3_2017_02_20-1.pdf
64 http://progress.unwomen.org/en/2015/pdf/UNW_progressreport.pdf
65 http://wbg.org.uk/wp-content/uploads/2016/11/De_Henau_WBG_childcare_briefing3_2017_02_20-1.pdf
66 Kim, Kijong and Antonopoulos, Rania (2011), 'Working Paper No. 691: Unpaid and Paid Care: The Effects of Child Care and Elder Care on the Standard of Living', Levy Economics Institute of Bard College

67 Ibid.

68 http://hdr.undp.org/sites/default/files/folbre_hdr_2015_final_0.pdf

69 http://newlaborforum.cuny.edu/2017/03/03/recognize-reduce-redistribute-unpaid-care-work-how-to-close-the-gender-gap/

70 http://wbg.org.uk/wp-content/uploads/2016/11/De_Henau_Perrons_WBG_CareEconomy_ITUC_briefing_final.pdf

71 http://newlaborforum.cuny.edu/2017/03/03/recognize-reduce-redistribute-unpaid-care-work-how-to-close-the-gender-gap/

72 http://wbg.org.uk/wp-content/uploads/2016/11/De_Henau_Perrons_WBG_CareEconomy_ITUC_briefing_final.pdf

73 Ibid.

74 http://wbg.org.uk/wp-content/uploads/2016/11/De_Henau_WBG_child-care_briefing3_2017_02_20-1.pdf

75 http://www.mckinsey.com/global-themes/gender-equality/the-power-of-parity-advancing-womens-equality-in-the-united-kingdom

76 http://wbg.org.uk/wp-content/uploads/2016/11/De_Henau_WBG_child-care_briefing3_2017_02_20-1.pdf

77 http://www.mckinsey.com/global-themes/gender-equality/the-power-of-parity-advancing-womens-equality-in-the-united-kingdom

78 http://wbg.org.uk/wp-content/uploads/2016/11/De_Henau_WBG_child-care_briefing3_2017_02_20-1.pdf

79 Ibid.

80 http://www.gothamgazette.com/city/6326-pre-k-offers-parents-opportunity-at-economic-gain

81 Ibid.

82 https://ourworldindata.org/women-in-the-labor-force-determinants

83 http://wbg.org.uk/wp-content/uploads/2016/11/De_Henau_WBG_child-care_briefing3_2017_02_20-1.pdf; http://progress.unwomen.org/en/2015/pdf/UNW_progressreport.pdf

84 http://www.mckinsey.com/global-themes/gender-equality/the-power-of-parity-advancing-womens-equality-in-the-united-kingdom http://wbg.org.uk/wp-content/uploads/2016/11/

85 http://wbg.org.uk/wp-content/uploads/2016/11/De_Henau_WBG_child-care_briefing3_2017_02_20-1.pdf

86 http://newlaborforum.cuny.edu/2017/03/03/recognize-reduce-redistribute-unpaid-care-work-how-to-close-the-gender-gap/

Chapter 13

1 https://twitter.com/alex6130/status/872937838488281088
2 https://twitter.com/MaliaBouattia/status/872978158135508992
3 https://twitter.com/DavidLammy/status/873063062483357696
4 https://www.buzzfeed.com/ikrd/we-dont-actually-know-how-many-young-people-turned-out-to?utm_term=.yw9j2lr8l#.cqOlx8Aa8
5 https://blog.oxforddictionaries.com/2017/12/14/youthquake-word-of-the-year-2017-commentary/
6 http://blogs.lse.ac.uk/politicsandpolicy/the-myth-of-the-2017-youth quake-election/
7 https://www.prospectmagazine.co.uk/blogs/peter-kellner/the-british-election-study-claims-there-was-no-youthquake-last-june-its-wrong
8 https://twitter.com/simonschusterUK/status/973882834665590785
9 https://oxfamblogs.org/fp2p/are-women-really-70-of-the-worlds-poor-how-do-we-know/;http://www.politifact.com/punditfact/article/2014/jul/03/meet-zombie-stat-just-wont-die/
10 https://www.americanprogress.org/issues/poverty/news/2013/03/11/56097/gender-equality-and-womens-empowerment-are-key-to-addressing-global-poverty/; https://www.theguardian.com/global-development-professionals-network/2013/mar/26/empower-women-end-poverty-developing-world; https://www.globalcitizen.org/en/content/introduction-to-the-challenges-of-achieving-gender/; https://www.pciglobal.org/womens-empowerment-poverty/; https://reliefweb.int/report/world/women-and-development-worlds-poorest-are-women-and-girls; http://www.ilo.org/global/about-the-ilo/newsroom/news/WCMS_008066/lang–en/index.htm; https://www.oecd.org/social/40881538.pdf
11 https://oxfamblogs.org/fp2p/are-women-really-70-of-the-worlds-poor-how-do-we-know/
12 http://www.politifact.com/punditfact/article/2014/jul/03/meet-zombie-stat-just-wont-die/
13 http://ideas4development.org/en/zombie-facts-to-bury-about-women-and-girls/
14 https://www.researchgate.net/profile/Rahul_Lahoti/publication/236248332_Moving_from_the_Household_to_the_Individual_Multi dimensional_Poverty_Analysis/links/5741941d08aea45ee8497aca/Moving-from-the-Household-to-the-Individual-Multidimensional-Poverty-Analysis.pdf?origin=publication_list
15 A note on household heads. The sex of a household head is determined by the sex of the other people living with her. A male-headed household can, and

often does, include adult females. A female-headed household, on the other hand, is almost always by definition a household that lacks an adult male – and lack is the operative term here. Households are not female-headed by default.

16 Lundberg, Shelly J., Pollak, Robert A. and Wales, Terence J. (1997), 'Do Husbands and Wives Pool Their Resources? Evidence from the United Kingdom Child Benefit', *Journal of Human Resources*, 32:3, 463–80, http://www.jstor.org/stable/146179

17 http://www.cpahq.org/cpahq/cpadocs/Feminization_of_Poverty.pdf; http://eprints.lse.ac.uk/3040/1/Gendered_nature_of_natural_disasters_%28LSERO%29.pdf

18 https://www.jstor.org/stable/145670?seq=1#page_scan_tab_contents; https://blogs.wsj.com/ideas-market/2011/01/27/the-gender-of-money/; François Bourguignon, Martin Browning, Pierre-André Chiappori and Valérie Lechene (1993), 'Intra Household Allocation of Consumption: A Model and Some Evidence from French Data', *Annales d'Économie et de Statistique*, 29, Progrès récents en théorie du consommateur / Recent Advances in Economic Theory, 137–56; http://jezebel.com/5744852/money-has-a-gender

19 https://www.theguardian.com/commentisfree/2017/may/01/conservatives-universal-credit-hard-work

20 https://docs.gatesfoundation.org/documents/gender-responsive-orientation-document.pdf

21 Ibid.

22 Gauff, Tonya Major (2009), 'Eliminating the Secondary Earner Bias: Lessons from Malaysia, the United Kingdom, and Ireland', *Northwestern Journal of Law and Social Policy*, 4:2

23 European Parliament (2017), *Gender Equality and Taxation in the European Union*, http://www.europarl.europa.eu/RegData/etudes/STUD/2017/583138/IPOL_STU%282017%29583138_EN.pdf

24 Andrienko, Yuri, Apps, Patricia and Rees, Ray (2014), 'Gender Bias in Tax Systems Based on Household Income', Discussion Paper, Institute for the Study of Labor

25 https://www.gov.uk/marriage-allowance/how-it-works

26 https://www.bloomberg.com/news/articles/2016–08-18/japan-may-finally-end-10–000-cap-on-women-s-incentive-to-work

27 http://www.undp.org/content/dam/undp/library/gender/Gender%20and%20Poverty%20Reduction/Taxation%20English.pdf

28 European Parliament (2017), *Gender Equality and Taxation in the European Union*, http://www.europarl.europa.eu/RegData/etudes/STUD/2017/583138/IPOL_STU%282017%29583138_EN.pdf
29 Ibid.
30 Ibid.
31 Institute of Development Studies (2016), 'Redistributing Unpaid Care Work – Why Tax Matters for Women's Rights'
32 https://wbg.org.uk/wp-content/uploads/2017/11/taxation-pre-Budget-nov-2017-final.pdf
33 Ibid.
34 Ibid.
35 Institute of Development Studies (2016), 'Redistributing Unpaid Care Work – Why Tax Matters for Women's Rights'
36 http://www.taxjustice.net/2016/11/03/switzerland-un-hot-seat-impact-tax-policies-womens-rights/
37 http://cesr.org/sites/default/files/downloads/switzerland_factsheet_2nov2016.pdf
38 http://www.undp.org/content/dam/undp/library/gender/Gender%20and%20Poverty%20Reduction/Taxation%20English.pdf 2010
39 Institute of Development Studies (2016), 'Redistributing Unpaid Care Work – Why Tax Matters for Women's Rights'
40 European Parliament (2017), *Gender Equality and Taxation in the European Union*

Chapter 14

1 https://www.politicalparity.org/wp-content/uploads/2015/08/Parity-Research-Women-Impact.pdf
2 http://www.historyandpolicy.org/policy-papers/papers/women-in-parliament-since-1945-have-they-changed-the-debate
3 https://www.diva-portal.org/smash/get/diva2:200156/FULLTEXT01.pdf
4 Australia, Austria, Belgium, Canada, Denmark, Finland, France, Greece, Ireland, Italy, the Netherlands, New Zealand, Norway, Portugal, Spain, Sweden, Switzerland, the United Kingdom and the United States
5 https://economics.mit.edu/files/792
6 https://web.stanford.edu/group/peg/Papers%20for%20call/nov05%20papers/Clots-Figueras.pdf
7 https://www.theatlantic.com/politics/archive/2016/09/clinton-trust-sexism/500489/

8 http://www.telegraph.co.uk/comment/3558075/Irrational-ambition-is-Hillary-Clintons-flaw.html
9 https://www.psychologytoday.com/blog/are-we-born-racist/201010/is-hillary-clinton-pathologically-ambitious
10 http://query.nytimes.com/gst/fullpage.html?res=9807E3D8123EF932A-15751C0A9619C8B63&sec=&spon=&pagewanted=2
11 http://www.weeklystandard.com/colin-powell-on-hillary-clinton-unbridled-ambition-greedy-not-transformational/article/2004328
12 http://www.teenvogue.com/story/hillary-clinton-laughs-too-ambitious-attack
13 http://www.dailymail.co.uk/news/article-3900744/Assange-says-Clinton-eaten-alive-ambitions-denies-Russia-Democratic-email-hacks-interview-Kremlin-s-TV-channel.html
14 http://www.theonion.com/blogpost/hillary-clinton-is-too-ambitious-to-be-the-first-f-11229
15 https://www.psychologytoday.com/blog/are-we-born-racist/201010/is-hillary-clinton-pathologically-ambitious
16 http://journals.sagepub.com/doi/10.1177/0146167210371949
17 http://www.sciencedirect.com/science/article/pii/S0022103108000334
18 Cikara, Mina and Fiske, Susan T. (2009), 'Warmth, competence, and ambivalent sexism: Vertical assault and collateral damage', in Barreto, Manuela, Ryan, Michelle K. and Schmitt, Michael T. (eds.), *The glass ceiling in the 21st century: Understanding barriers to gender equality*, Washington
19 https://www.sciencedaily.com/releases/2016/08/160829095050.htm
20 Hekman, David, Johnson, Stefanie, Foo, Maw-Der and Yang, Wei (2017), 'Does Diversity-Valuing Behavior Result in Diminished Performance Ratings for Non-White and Female Leaders?', *Academy of Management Journal*, 60:2, 771
21 https://www.lrb.co.uk/v39/n02/rebecca-solnit/from-lying-to-leering
22 http://archive.ipu.org/wmn-e/world.htm
23 https://www.parliament.uk/business/committees/committees-a-z/commons-select/women-and-equalities-committee/news-parliament-2017/govt-response-women-hoc-2017–19/
24 https://www.gov.uk/government/uploads/system/uploads/attachment_data/file/642904/Government_Response_-_Women_in_the_House_of_Commons.pdf
25 http://archive.ipu.org/wmn-e/arc/classif010197.htm
26 https://www.fawcettsociety.org.uk/Handlers/Download.ashx?IDMF=2e149e34-9c26-4984-bf64-8989db41a6ad

27 https://www.gov.uk/government/uploads/system/uploads/attachment_data/file/642904/Government_Response_-_Women_in_the_House_of_Commons.pdf
28 Diana Z. O'Brien and Johanna Rickne (2016), 'Gender Quotas and Women's Political Leadership', *American Political Science Review*, 110:1 (February 2016), 112–26
29 https://blogs.eui.eu/genderquotas/wp-content/uploads/sites/24/2015/03/Executive-summary-Sweden-Freidenvall1.pdf
30 Ibid.
31 http://www.europarl.europa.eu/RegData/etudes/note/join/2013/493011/IPOL-FEMM_NT(2013)493011_EN.pdf
32 https://blogs.eui.eu/genderquotas/wp-content/uploads/sites/24/2015/03/Executive-summary-Sweden-Freidenvall1.pdf
33 Yoon, J. and Shin, K. (2015), 'Mixed effects of legislative quotas in South Korea', *Politics & Gender*, 11:1, 186–95
34 O'Brien and Rickne (2016)
35 https://web.archive.org/web/20110605021810/http://www.parliament.the-stationery-office.co.uk/pa/cm200708/cmhansrd/cm080306/debtext/80306-0007.htm
36 https://www.worksopguardian.co.uk/news/politics/man-who-sent-mp-s-wife-dead-bird-in-post-is-given-restraining-order-1-4777574
37 https://christinescottcheng.wordpress.com/publications/women-in-politics/mixed-member-proportional-leads-to-more-women-mps/how-the-electoral-system-matters-for-electing-women/; http://www.europarl.europa.eu/working papers/femm/w10/2_en.htm
38 Castillejo, Clare (2016), 'Women political leaders and peacebuilding', http://noref.no/var/ezflow_site/storage/original/application/6ccaf3f24b120b8004f0db2a767a9dc2.pdf
39 http://reliefweb.int/sites/reliefweb.int/files/resources/6ccaf3f24b120b8004f0db2a767a9dc2.pdf
40 Castillejo (2016)
41 http://www.capwip.org/readingroom/cawp-womenstateleg.pdf
42 https://www.fawcettsociety.org.uk/Handlers/Download.ashx?IDMF=2e149e34-9c26-4984-bf64-8989db41a6ad
43 Hancock, Adrienne B., Rubin, Benjamin A. (2015), 'Influence of Communication Partner's Gender on Language', *Journal of Language and Social Psychology*, 34:1, 46–64
44 http://www.pbs.org/newshour/rundown/for-many-women-watching-trump-interrupt-clinton-51-times-was-unnerving-but-familiar/

45 https://www.vanityfair.com/news/2017/11/inside-the-fall-of-todays-matt-lauer
46 https://hbr.org/2016/09/why-hillary-clinton-gets-interrupted-more-than-donald-trump
47 https://www.theguardian.com/commentisfree/2016/mar/28/hillary-clinton-honest-transparency-jill-abramson
48 http://www.bbc.co.uk/news/uk-politics-13211577
49 http://archive.ipu.org/pdf/publications/issuesbrief-e.pdf
50 https://www.theguardian.com/politics/2014/feb/09/fawzia-koofi-afghanistan-mp-turn-off-microphones
51 https://www.theguardian.com/technology/datablog/ng-interactive/2016/jun/27/from-julia-gillard-to-hillary-clinton-online-abuse-of-politicians-around-the-world
52 http://archive.ipu.org/pdf/publications/issuesbrief-e.pdf
53 Ibid.
54 http://www.medicamondiale.org/fileadmin/redaktion/5_Service/Mediathek/Dokumente/English/Documentations_studies/medica_mondiale_-_Report_on_Women__Peace_and_Security_-_October_2007.pdf
55 https://www.theguardian.com/politics/2014/feb/09/fawzia-koofi-afghanistan-mp-turn-off-microphones
56 https://www.reuters.com/article/us-afghanistan-women/bomb-attack-in-eastern-afghanistan-kills-female-politician-idUSKBN0LK1EI20150216
57 O'Brien and Rickne (2016)
58 Ibid.
59 http://archive.ipu.org/wmn-e/classif.htm
60 Kanthak, Kristin and Krause, George A. (2012), *The Diversity Paradox: Political Parties, Legislatures, and the Organizational Foundations of Representation in America*, New York http://www.pitt.edu/~gkrause/Kanthak%20&%20Krause.Diversity%20Paradox.Book%20Manuscript.09–10-10.pdf
61 O'Brien and Rickne (2016)
62 Kanthak and Krause (2012) – once women threaten the majority, there is a male backlash
63 Wittmer, Dana and Bouche, Vanessa (2010), 'The Limits of Gendered Leadership: The Public Policy Implications of Female Leadership on Women's Issues', *The Limits of Gendered Issues*, APSA 2010 Annual Meeting Paper
64 http://archive.ipu.org/pdf/publications/issuesbrief-e.pdf
65 http://www.medicamondiale.org/fileadmin/redaktion/5_Service/Mediathek/Dokumente/English/Documentations_studies/medica_mondiale_-_Report_on_Women__Peace_and_Security_-_October_2007.pdf

66 http://archive.ipu.org/pdf/publications/issuesbrief-e.pdf
67 https://www.cfr.org/article/violence-against-female-politicians
68 http://archive.ipu.org/pdf/publications/issuesbrief-e.pdf
69 https://www.cfr.org/article/violence-against-female-politicians
70 https://www.ndi.org/sites/default/files/not-the-cost-program-guidance-final.pdf
71 https://www.ndi.org/sites/default/files/not-the-cost-program-guidance-final.pdf
72 https://www.cfr.org/article/violence-against-female-politicians
73 Jacobi, Tonja and Schweers, Dylan (2017), 'Justice, Interrupted: The Effect of Gender, Ideology and Seniority at Supreme Court Oral Arguments' (14 March 2017), *Virginia Law Review*, 1379, *Northwestern Law & Econ Research Paper No. 17–03*
74 http://www.bbc.com/capital/story/20170622-why-women-should-interrupt-men
75 http://www.bbc.com/capital/story/20160906-how-rude-the-secret-to-smart-interrupting
76 https://www.nytimes.com/2017/06/13/us/politics/kamala-harris-interrupted-jeff-sessions.html
77 http://edition.cnn.com/2017/06/13/politics/powers-miller-kamala-harris-hysterical-sessions-hearing-ac360-cnntv/index.html
78 http://interactions.acm.org/archive/view/january-february-2014/are-you-sure-your-software-is-gender-neutral
79 Tali Mendelberg, a Princeton University professor who is co-author of *The Silent Sex: Gender, Deliberation and Institutions*, https://mobile.nytimes.com/2016/10/27/upshot/speaking-while-female-and-at-a-disadvantage.html?em_pos=small&emc=edit_up_20161028&nl=upshot&nl_art=3&nlid=67555443&ref=headline&te=1&_r=0&referer=http://m.facebook.com
80 http://time.com/3666135/sheryl-sandberg-talking-while-female-manterruptions/
81 Mendelberg
82 Karpowitz, C., Mendelberg, T. and Shaker, L. (2012) 'Gender Inequality in Deliberative Participation', *American Political Science Review*, 106:3, 533–47
83 https://www.fawcettsociety.org.uk/Handlers/Download.ashx?IDMF=2e149e34-9c26-4984-bf64-8989db41a6ad
84 https://www.bindmans.com/insight/updates/when-can-the-law-remove-a-councillor-without-an-election;http://localgovernmentlawyer.co.uk/index.php?option=com_content&view=article&id=17463%3Acouncil-blames-local-

ism-act-for-inability-to-remove-councillor-from-office&catid=59%3Agovernance-a-risk-articles&Itemid=27

85 https://www.fawcettsociety.org.uk/Handlers/Download.ashx?IDMF=2e149e34-9c26-4984-bf64-8989db41a6ad

Chapter 15

1 http://www.makers.com/once-and-for-all

2 https://www.globalfundforwomen.org/wp-content/uploads/2006/11/disaster-report.pdf

3 https://www.womensrefugeecommission.org/gbv/firewood

4 http://gdnonline.org/resources/women_will_rebuild_miami.pdf

5 Murakami-Ramalho, E. and Durodoye, B. (2008), 'Looking Back to Move Forward: Katrina's Black Women Survivors Speak', *NWSA Journal*, 20(3), 115–37

6 https://iwpr.org/wp-content/uploads/wpallimport/files/iwpr-export/publications/D506_GetToTheBricks.pdf

7 https://www.theguardian.com/global-development/2015/jan/22/women-rights-war-peace-un-resolution-1325

8 http://www.peacewomen.org/assets/file/NationalActionPlans/miladpournikanalysisdocs/igis_womeninpeaceandsecuritythroughunsr1325_millerpournikswaine_2014.pdf; Data2x

9 https://www.cfr.org/interactive/interactive/womens-participation-in-peace-processes/explore-the-data

10 https://reliefweb.int/sites/reliefweb.int/files/resources/UNW-GLOBAL-STUDY-1325–2015.pdf

11 United Nations Security Council (2017) 'Report of the Secretary-General on women and peace and security'

12 Clare Castillejo (2016), 'Women political leaders and peacebuilding', Norwegian Peacebuilding Resource Centre

13 Ibid.

14 http://www.unwomen.org/en/what-we-do/peace-and-security/facts-and-figures#sthash.vq3NnLEu.dpuf; O'Reilly, Marie, Súilleabháin, Andrea Ó and Paffenholz, Thania (2015), 'Reimagining Peacemaking: Women's Roles in Peace Processes', International Peace Institute, New York

15 Castillejo (2016)

16 O'Reilly, Súilleabháin and Paffenholz (2015)

Chapter 16

1 http://www.securitycouncilreport.org/atf/cf/%7B65BFCF9B-6D27-4E9C-8CD3-CF6E4FF96FF9%7D/WPS%202010%20Sidebar2.pdf

2 http://www.un.org/en/preventgenocide/rwanda/about/bgsexualviolence.shtml

3 O'Reilly, Marie, Súilleabháin, Andrea Ó and Paffenholz, Thania (2015), 'Reimagining Peacemaking: Women's Roles in Peace Processes,' International Peace Institute, New York, June 2015, https://www.ipinst.org/wp-content/uploads/2015/06/IPI-E-pub-Reimagining-Peacemaking.pdf

4 http://www.un.org/en/preventgenocide/rwanda/about/bgsexualviolence.shtml

5 O'Reilly, Súilleabháin and Paffenholz (2015)

6 http://www.unwomen.org/en/what-we-do/peace-and-security/facts-and-figures#sthash.vq3NnLEu.dpuf

7 http://www.nytimes.com/2013/12/11/world/asia/effort-to-help-filipino-women-falters-un-says.html

8 http://www.unfpa.org/press/women%E2%80%99s-health-critical-recovery-aftermath-typhoon-haiyan-says-unfpa-executive-director

9 Ibid.

10 http://www.nytimes.com/2013/12/11/world/asia/effort-to-help-filipino-women-falters-un-says.html

11 O'Reilly, Súilleabháin and Paffenholz (2015)

12 http://www.indexmundi.com/facts/oecd-members/maternal-mortality-ratio

13 https://www.unicef.org/childsurvival/sierraleone_91206.html

14 http://www.thesierraleonetelegraph.com/?p=16261

15 https://www.washingtonpost.com/national/health-science/2014/08/14/3e08d0c8-2312-11e4-8593-da634b334390_story.html?utm_term=.51eb39dc57dc

16 Ibid.

17 http://www.thelancet.com/journals/langlo/article/PIIS2214-109X(15)00065–0/fulltext

18 http://www.wpro.who.int/topics/gender_issues/Takingsexandgenderintoaccount.pdf

19 http://theconversation.com/zika-and-ebola-had-a-much-worse-effect-on-women-we-need-more-research-to-address-this-in-future-64868

20 Ibid.

21 http://foreignpolicy.com/2014/08/20/why-are-so-many-women-dying-from-ebola/

22 http://www.unwomen.org/en/news/stories/2016/5/ed-statement-on-whs#sthash.xmKEsOBX.dpuf

23 https://www.washingtonpost.com/national/health-science/2014/08/14/3e08d0c8-2312-11e4-8593-da634b334390_story.html?utm_term=.51eb39dc57dc

24 https://www.buzzfeed.com/jinamoore/ebola-is-killing-women-in-far-greater-numbers-than-men?utm_term=.gpzKwwzJze#.wce6ww292m

25 Ibid.

26 https://www.washingtonpost.com/national/health-science/2014/08/14/3e08d0c8-2312-11e4-8593-da634b334390_story.html?utm_term=.51eb39dc57dc

27 https://www.chathamhouse.org/publication/ia/gendered-human-rights-analysis-ebola-and-zika-locating-gender-global-health# https://www.chathamhouse.org/publication/ia/gendered-human-rights-analysis-ebola-and-zika-locating-gender-global-health

28 http://theconversation.com/zika-and-ebola-had-a-much-worse-effect-on-women-we-need-more-research-to-address-this-in-future-64868

29 https://www.theguardian.com/environment/blog/2014/jul/14/8-charts-climate-change-world-more-dangerous

30 https://www.washingtonpost.com/news/energy-environment/wp/2016/07/25/how-climate-disasters-can-drive-violent-conflict-around-the-world/?utm_term=.8b5c33ad65e7

31 https://www.washingtonpost.com/opinions/another-deadly-consequence-of-climate-change-the-spread-of-dangerous-diseases/2017/05/30/fd3b8504-34b1-11e7-b4ee-434b6d506b37_story.html?utm_term=.e49b6bd86143

32 http://www.thelancet.com/journals/lanplh/article/PIIS2542-5196(17)30082-7/fulltext?elscal=tlpr

33 http://edition.cnn.com/2017/08/04/health/climate-change-weather-disasters-europe/index.html

34 Neumayer, Eric and Plümper, Thomas (2007) 'The gendered nature of natural disasters: the impact of catastrophic events on the gender gap in life expectancy, 1981–2002', *Annals of the Association of American Geographers*, 97:3, 55–66

35 Ibid.

36 https://www.theguardian.com/society/2005/mar/26/internationalaidand development.indianoceantsunamidecember2004
37 http://eprints.lse.ac.uk/3040/1/Gendered_nature_of_natural_disasters_%28LSERO%29.pdf
38 Ibid.
39 https://thewire.in/66576/economic-growth-bangladesh-challenge-change-women/
40 http://www.bridge.ids.ac.uk/sites/bridge.ids.ac.uk/files/reports/Climate_Change_DFID.pdf
41 https://www.globalfundforwomen.org/wp-content/uploads/2006/11/disaster-report.pdf
42 Ibid.
43 https://iwpr.org/wp-content/uploads/wpallimport/files/iwpr-export/publications/D506_GetToTheBricks.pdf
44 https://qz.com/692711/the-radically-simple-way-to-make-female-refugees-safer-from-sexual-assault-decent-bathrooms/
45 https://www.womensrefugeecommission.org/facts-and-figures
46 https://www.globalone.org.uk/wp-content/uploads/2017/03/SYRIA-REPORT-FINAL-ONLINE.pdf
47 https://www.amnesty.ie/greece-refugee-women-coping-fear-violence-camps/ https://blogs.cdc.gov/global/2014/11/17/implications-of-latrines-on-womens-and-girls-safety/
48 http://refugeerights.org.uk/wp-content/uploads/2017/03/RRDP_Hidden-Struggles.pdf
49 http://www.wame2015.org/case-study/1124/ https://blogs.cdc.gov/global/2014/11/17/implications-of-latrines-on-womens-and-girls-safety/
50 https://www.hrw.org/sites/default/files/report_pdf/wrdsanitation0417_web_0.pdf
51 http://refugeerights.org.uk/wp-content/uploads/2017/03/RRDP_Hidden-Struggles.pdf
52 https://qz.com/692711/the-radically-simple-way-to-make-female-refugees-safer-from-sexual-assault-decent-bathrooms/
53 https://www.theguardian.com/lifeandstyle/2018/jul/09/i-couldnt-even-wash-after-giving-birth-how-showers-are-restoring-the-dignity-of-female-refugees
54 IRIN and women's refugee council
55 https://www.hrw.org/sites/default/files/report_pdf/wrdsanitation0417_web_0.pdf
56 Ibid.

57 http://www.bbc.co.uk/news/uk-england-beds-bucks-herts-36804714

58 https://www.irinnews.org/investigations/2017/05/10/women-refugees-risk-sexual-assault-berlin-shelters

59 https://www.buzzfeed.com/jinamoore/women-refugees-fleeing-through-europe-are-told-rape-is-not-a?utm_term=.lmMmNv6vBq#.tgnEGvDv89

60 https://www.theguardian.com/world/2017/feb/28/refugee-women-and-children-beaten-raped-and-starved-in-libyan-hellholes

61 https://www.irinnews.org/investigations/2017/05/10/women-refugees-risk-sexual-assault-berlin-shelters

62 Ibid.

63 http://www.huffingtonpost.com/soraya-chemaly/women-and-disaster-relief_b_5697868.html

64 http://odihpn.org/magazine/linking-food-security-food-assistance-and-protection-from-gender-based-violence-wfp%C2%92s-experience/

65 https://www.telegraph.co.uk/news/2018/02/17/oxfam-warned-decade-ago-crisis-sex-abuse-among-worlds-aid-workers/

66 https://wssagwales.files.wordpress.com/2012/10/asawoman.pdf

67 https://phys.org/news/2017–04-uk-hidden-homeless-lone-women.html

68 https://www.theguardian.com/society/2017/dec/14/homelessness-women-seeking-support-outnumber-men-for-first-time?CMP=share_btn_tw

69 https://phys.org/news/2017–04-uk-hidden-homeless-lone-women.html

70 http://www.feantsa.org/download/feantsa-ejh-11–1_a1-v045913941269604492255.pdf

71 https://www.policyalternatives.ca/publications/commentary/fast-facts-4-things-know-about-women-and-homelessness-canada

72 http://www.feantsa.org/download/feantsa-ejh-11–1_a1-v045913941269604492255.pdf

73 https://phys.org/news/2017–04-uk-hidden-homeless-lone-women.html

74 https://www.theguardian.com/commentisfree/2017/apr/19/sex-rent-logical-extension-leaving-housing-to-market

75 https://www.policyalternatives.ca/publications/commentary/fast-facts-4-things-know-about-women-and-homelessness-canada

76 http://journals.sagepub.com/doi/abs/10.1177/088626001016008001

77 https://www.bustle.com/articles/190092-this-is-how-homeless-women-cope-with-their-periods

78 https://www.theguardian.com/housing-network/2016/aug/22/sex-in-return-for-shelter-homeless-women-face-desperate-choices-government-theresa-may

79 https://www.telegraph.co.uk/women/womens-health/11508497/TheHomelessPeriod-Campaign-for-homeless-women-to-have-free-tampons.html
80 http://thehomelessperiod.com/
81 https://www.change.org/p/help-the-homeless-on-their-period-thehomeless-period/u/19773587
82 https://www.thecut.com/2016/06/nyc-will-provide-tampons-in-schools-shelters.html
83 http://www.unhcr.org/uk/news/latest/2008/4/4815db792/corporate-gift-highlights-sanitation-problems-faced-female-refugees.html; http://www.reuters.com/article/us-womens-day-refugees-periods-feature-idUSKBN16F1UU
84 https://www.womensrefugeecommission.org/images/zdocs/Refugee-Women-on-the-European-Route.pdf; https://www.globalone.org.uk/wp-content/uploads/2017/03/SYRIA-REPORT-FINAL-ONLINE.pdf; https://globalone.org.uk/2017/05/a-14-year-olds-heart-wrenching-tale/
85 http://www.ifrc.org/en/news-and-media/news-stories/africa/burundi/upholding-women-and-girls-dignity-managing-menstrual-hygiene-in-emergency-situations-62536/
86 January 2016, https://www.womensrefugeecommission.org/images/zdocs/Refugee-Women-on-the-European-Route.pdf; http://www.nytimes.com/2013/12/11/world/asia/effort-to-help-filipino-women-falters-un-says.html
87 http://www.ifrc.org/en/news-and-media/news-stories/africa/burundi/upholding-women-and-girls-dignity-managing-menstrual-hygiene-in-emergency-situations-62536/
88 http://www.reuters.com/article/us-womens-day-refugees-periods-feature-idUSKBN16F1UU
89 https://www.globalone.org.uk/wp-content/uploads/2017/03/SYRIA-REPORT-FINAL-ONLINE.pdf

Afterword

1 http://discovermagazine.com/2006/mar/knit-theory
2 http://www.cabinetmagazine.org/issues/16/crocheting.php
3 https://www.brainpickings.org/2009/04/24/margaret-wertheim-institute-for-figuring/
4 http://discovermagazine.com/2006/mar/knit-theory
5 Ibid.

6 http://www.cabinetmagazine.org/issues/16/crocheting.php

7 https://www.nytimes.com/2005/07/11/nyregion/professor-lets-her-fingers-do-the-talking.html

8 Cikara, Mina and Fiske, Susan T. (2009), 'Warmth, competence, and ambivalent sexism: Vertical assault and collateral damage', in Barreto, Manuela, Ryan, Michelle K. and Schmitt, Michael T. (eds.), *The glass ceiling in the 21st century: Understanding barriers to gender equality*, Washington DC

9 https://link.springer.com/content/pdf/10.1007%2Fs00004-000-0015-0.pdf

10 https://www.wired.com/2008/06/pb-theory/

11 http://womenintvfilm.sdsu.edu/wp-content/uploads/2017/01/2016_Celluloid_Ceiling_Report.pdf

12 http://wmc.3cdn.net/dcdb0bcb4b0283f501_mlbres23x.pdf

13 http://www.slate.com/articles/news_and_politics/history/2016/01/popular_history_why_are_so_many_history_books_about_men_by_men.html?via=gdpr-consent#methodology

14 https://www.historians.org/publications-and-directories/perspectives-on-history/may-2010/what-the-data-reveals-about-women-historians

15 https://www.historians.org/publications-and-directories/perspectives-on-history/december-2015/the-rise-and-decline-of-history-specializations-over-the-past-40-years

16 http://duckofminerva.com/2015/08/new-evidence-on-gender-bias-in-ir-syllabi.html

17 https://www.theguardian.com/world/2004/jan/23/gender.uk

18 *Sex Discrimination Law Review* (January 2018), www.fawcettsociety.org.uk

19 Nielsen, Mathias Wullum, Andersen, Jens Peter, Schiebinger, Londa and Schneider, Jesper W. (2017), 'One and a half million medical papers reveal a link between author gender and attention to gender and sex analysis', *Nature Human Behaviour*, 1, 791–6

20 https://www.vox.com/policy-and-politics/2017/7/18/15991020/3-gop-women-tank-obamacare-repeal

21 Ransby, B. (2006), 'Katrina, Black Women, and the Deadly Discourse on Black Poverty in America', *Du Bois Review: Social Science Research on Race*, 3:1, 215–22, DOI:10.1017/S1742058X06060140

22 https://grist.org/article/hurricane-maria-hit-women-in-puerto-rico-the-hardest-and-theyre-the-ones-building-it-back/

23 https://www.vogue.com/projects/13542078/puerto-rico-after-hurricane-maria-2/

Epilogue

1 https://blogs.bmj.com/bmjgh/2020/03/24/sex-gender-and-covid-19-disaggregated-data-and-health-disparities/?utm_campaign=Invisible%20Women&utm_medium=email&utm_source=Revue%20newsletter

2 https://www.devex.com/news/opinion-we-lack-an-essential-component-to-power-covid-19-response-98054

3 https://www.nytimes.com/2020/04/27/health/coronavirus-estrogen-men.html

4 https://www.getrevue.co/profile/InvisibleWomen/issues/invisible-women-default-male-hangovers-274639

5 https://www.medrxiv.org/content/10.1101/2020.09.13.20193680v1?%253fcollection=

6 https://www.cnbc.com/2020/03/18/how-the-coronavirus-could-impact-women-in-health-care.html

7 https://www.cnbc.com/2020/03/18/how-the-coronavirus-could-impact-women-in-health-care.html

8 https://phinational.org/wp-content/uploads/legacy/phi-home-care-workers-key-facts.pdf

9 https://www.cam.ac.uk/research/news/women-bear-brunt-of-coronavirus-economic-shutdown-in-uk-and-us; https://www.theguardian.com/world/2020/may/24/pink-collar-recession-how-the-covid-19-crisis-is-eroding-womens-economic-power; https://www.theguardian.com/australia-news/audio/2020/may/12/how-the-coronavirus-is-affecting-womens-work-in-australia; https://www.theguardian.com/world/2020/mar/29/low-paid-women-in-uk-at-high-risk-of-coronavirus-exposure; https://www.vox.com/2020/4/23/21229942/coronavirus-grocery-store-workers-walmart-covid-pandemic; https://www.dw.com/en/coronavirus-disproportionately-affects-womens-health-finances/a-53319007; https://www.linkedin.com/pulse/covid-19-impacting-womens-employment-opportunities-janine-chamberlin/?trackingId=9fLBOpnQQqaIkSHcOU86dw%3D%3D; https://www.nber.org/papers/w26947.pdf

10 https://www.nytimes.com/2020/05/06/upshot/pandemic-chores-homeschooling-gender.html; https://www.bbc.co.uk/news/business-52808930; https://www.theguardian.com/world/2020/may/27/working-mothers-interrupted-more-often-than-fathers-in-lockdown-study; https://www.thetimes.co.uk/edition/news/coronavirus-loading-mothers-with-washing-cooking-and-home-schooling-2spg29b3n; https://www.ifs.org.uk/

publications/14860; https://www.theguardian.com/inequality/2020/jun/18/uk-society-regressing-back-to-1950s-for-many-women-warn-experts-worsening-inequality-lockdown-childcare

11 https://www.protocol.com/parents-working-from-home-survey?utm_campaign=Invisible%20Women&utm_medium=email&utm_source=Revue%20newsletter

12 https://www.bcg.com/publications/2020/valuable-productivity-gains-covid-19?utm_campaign=Invisible%20Women&utm_medium=email&utm_source=Revue%20newsletter

13 https://www.gov.uk/government/news/employers-do-not-have-to-report-gender-pay-gaps#:~:text=Due%20to%20the%20Coronavirus%20outbreak,year%20(2019%2F20).

14 https://www.theguardian.com/money/2020/jul/24/uk-working-mothers-are-sacrifical-lambs-in-coronavirus-childcare-crisis; https://www.theatlantic.com/ideas/archive/2020/07/reopening-bars-easy-schools-are-difficult/613861/; https://www.thenation.com/article/society/economy-reopen-childcare/; https://www.theguardian.com/money/2020/sep/13/reopening-uk-offices-risks-excluding-women-and-minorities-says-business-chief?utm_term=Autofeed&CMP=twt_b-gdnnews&utm_medium=Social&utm_source=Twitter#Echobox=1600011463; https://news.bloomberglaw.com/daily-labor-report/without-child-care-back-to-work-parents-have-few-legal-options

15 WBG report

16 https://www.bloomberg.com/news/articles/2020-06-29/johnson-promises-new-deal-to-rebuild-u-k-economy-after-virus?sref=pSQ7fiVq

Index